Sedimentary Structures in Dunes of the Namib Desert, South West Africa

Edwin D. McKee
U.S. Geological Survey
M.S. 921, Federal Center
Denver, Colorado 80225

SPECIAL PAPER

188

THE GEOLOGICAL SOCIETY OF AMERICA
P.O. Box 9140 · 3300 Penrose Place · Boulder, Colorado 80301

Published by The Geological Society of America, Inc.,
P.O. Box 9140, 3300 Penrose Place, Boulder, Colorado 80301

Printed in U.S.A.

Library of Congress Cataloging in Publication Data

McKee, Edwin Dinwiddie, 1906—
Sedimentary structures in dunes of the Namib Desert,
South West Africa.

(Special paper / the Geological Society of America;
188)
Bibliography: p.
1. Sand-dunes—Namibia—Namib Desert. I. Title.
II. Series: Special paper (Geological Society of
America); 188.
GB638.88.N35M37 551.3′75 81-20155
ISBN 0-8137-2188-1 AACR2

Contents

Geological Society of America
Special Paper 188
1982

Sedimentary Structures in Dunes of the Namib Desert, South West Africa

Abstract

The sand sea that composes a large part of the Namib Desert in South West Africa covers approximately 34,000 km^2 and extends from the Great Escarpment on the east to the Atlantic Ocean on the west, south of the Kuiseb River. This sand sea, or erg, contains most of the principal dune types that have been recognized throughout the world, many of them of great height and separated by broad, flat interdunes. The dominant type and by far the most common form is the linear (seif, longitudinal) dune that extends as a series of long, nearly parallel ridges with north-south orientation across the length of the dune field.

Studies of dune structure were made largely by trenching the various dune types to expose sections of cross-strata in selected parts. The trenches showed that those dunes attributed to unidirectional winds—barchan, barchanoid ridge, and transverse—with low-angle foresets on the windward sides and high angles on the lee sides, are mostly located along a narrow belt on the Atlantic Coast.

The large interior dunes of linear type showed high-angle dips on both sides of the ridges and are believed to result from alternating two-directional winds. Because the genesis of these dunes has long been controversial, a review is given of the other principal hypotheses—deflation, helical roll, modification of simple dune forms, and outcrop-controlled lee-side accumulation—and evidence is presented for and against each.

The most complex of dune types, in which arms radiate from a central point, is referred to as a star dune and is fairly numerous in various parts of the sand sea, especially near the vleis and along the northern dune margin. Such dunes are considered, both on the basis of data from trenching and from records of measured winds, to be the product of wind from three or more directions. Star dunes form along the ridges of linear dunes and also as isolated mounds on the desert floor, and they attain considerable heights.

Minor dune types that are described and briefly discussed are those controlled by vegetation and referred to as coppice dunes. Also the small types known as dome dunes and blowouts, both of which were structurally analyzed, are discussed.

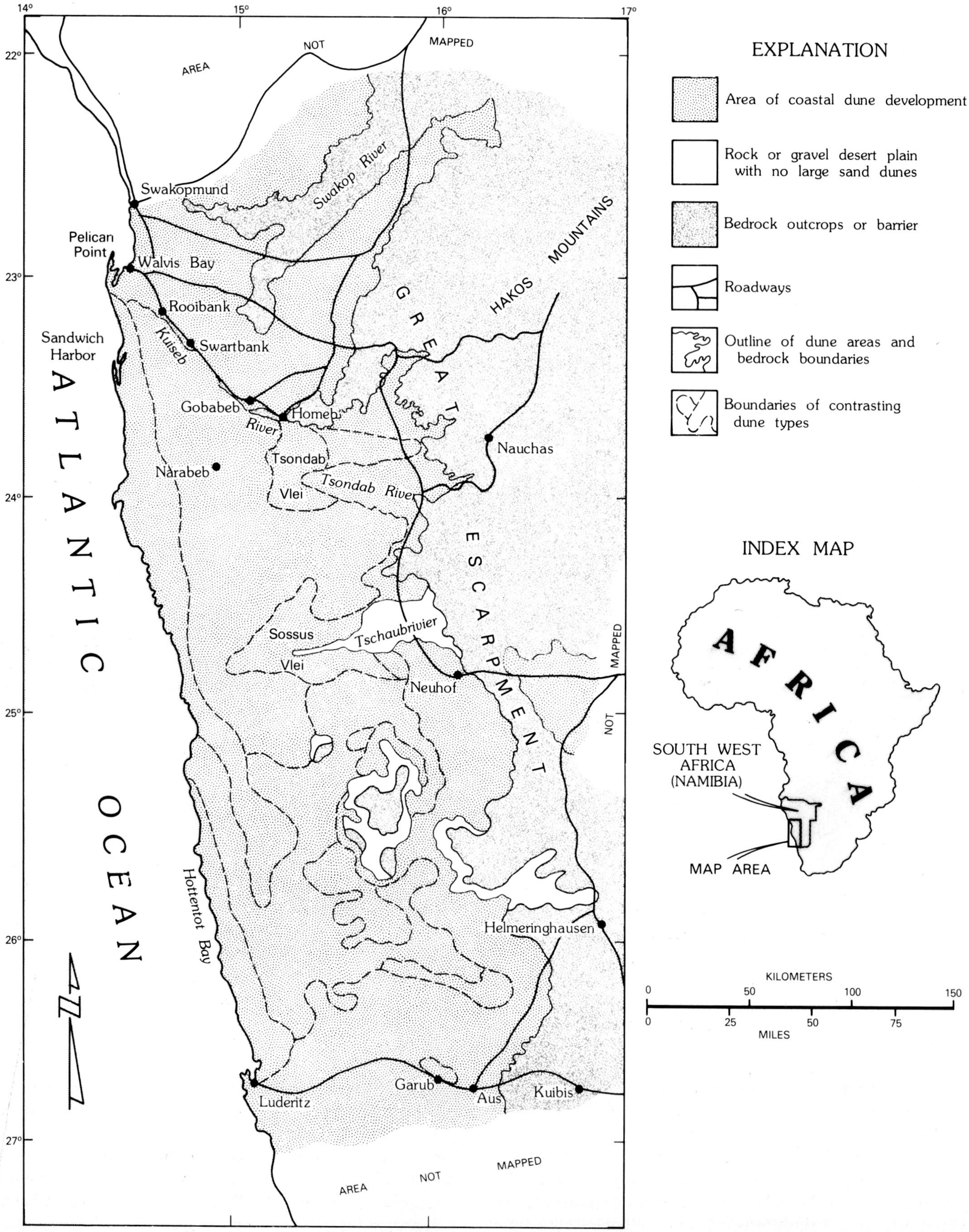

Figure 1. Map of Namib Desert sand sea, South West Africa (Namibia).

Introduction

Objectives

The principal purpose of this report is to describe the structures of dunes and interdunes within the sand sea of the Namib Desert in South West Africa (Namibia) and to demonstrate the value of these structures in interpreting the genesis and history of the dune field. The Namib sand sea, or erg, is a mass of sand that is great both from the standpoint of areal extent and from the height and bulk of individual dunes. The dune field (Fig. 1) was described by Barnard (1973, p. 2) as covering 34,000 km^2; its width from the Atlantic Coast in the west to the Great Escarpment in the east ranges from 45 to 130 km, and it extends from the area near Aus and Luderitz (26° 38′ S.) northward to the Kuiseb River (22° 51′ S.) and slightly beyond—approximately 400 km in length (Nagtegaal, 1973, p. 7).

Climate

The climate of the central Namib Desert—the area occupied by the sand sea—has been described by many people (Logan, 1960; Goudie, 1972; Besler, 1976; and others) so will not be discussed in detail here. Research in this field is a continuing project of the Desert Ecological Research Unit, and meteorological data are collected on a regular schedule at the weather station at Gobabeb. Seven stations between the Atlantic Coast and the eastern edge of the desert, 100 km inland, record precipitation, humidity, wind, and temperature. A brief summary of the principal elements of the climate follows.

Precipitation

Rainfall in the Namib sand sea is extremely low; average annual totals, as given by Goudie (1972, p. 16), range from 15 mm at Luderitz in the south and 14 mm at Swakopmund in the north to 23 mm at Gobabeb in the interior. The moisture-laden air responsible for the rain has been shown to have come from the Indian Ocean to the east and to be operative largely during the summer (Logan, 1960, p. 136; Scholz, 1972, p. 33). The annual rainfall decreases steadily from the Great Escarpment on the east toward the Atlantic Ocean (Goudie, 1972, p. 16).

Moisture in the form of fog represents a second type of precipitation and contributes substantially to the total amount, especially in coastal areas. As described by Goudie (1972, p. 16), "Whilst the fog only precipitated on an average of 60 days in the year, this produced a precipitation of 31 mm per annum, a figure slightly higher than the rainfall itself" (23 mm per annum). The effect of this fog on the form and migration of dunes probably is appreciable.

Temperatures

The most striking feature of temperature in the Namid Desert is the similarity in average monthly and yearly measurements, as shown in a table compiled for monthly means at Walvis Bay, Luderitz, and Gobabeb (Goudie, 1972, p. 17, Table 1). The mean and annual range at two stations between warmest and coldest months is said to be 9 °C, and the annual range at Gobabeb is given as only 6.5 °C.

Wind Distribution

The general pattern of wind circulation in the desert areas of South West Africa has been recognized for a considerable time and has been described by many writers. It involves two principal systems, as stated by Logan (1960, p. 136) and by Goudie (1972, p. 24): (1) a daily southwesterly sea breeze blowing off the South Atlantic toward the northeast and (2) easterly winds of high velocity called berg winds which, though less frequent, are more violent. Lesser winds from the northwest are also recorded for short seasons. Further details of this wind pattern and a map of sand roses from the area are presented in the section of this report on Wind Regimes Associated with Namib Linear Dunes under Dune Types.

Information on wind patterns in the vicinity of Gobabeb, presented in tabular form by Goudie (1972, Table 2), shows that the distribution of desert sand-moving winds is from three, rather than two, principal directions. The easterly and northeasterly winds, in terms of hour frequency, include a majority of all the high-velocity measurements (35+ km/hour); the westerly and southwesterly winds show a preponderance of the low-velocity winds (20 to 29 km/hour) that also move sands. In addition to these domi-

A

B

Figure 2 (this and facing page). Interdune sand deposits of interior Namib Desert. *A*, sand surface, 2 km south of Homeb; *B*, gravelly sand surface, 12 km south of Gobabeb; *C*, wind-rippled sand and vegetation mounds, Soutrivier area, southwest of Gobabeb; *D*, shrub and grass coverage, sand surface 5 km south of Gobabeb. Photographs by E. Tad Nichols.

C

D

nant winds, others from the north and northwest include a small but appreciable number of hours in both low- and high-velocity categories.

A measure of the sand-moving capability of winds in the Gobabeb area over a 4-year period has been determined by Besler (1975, p. 135). Dune fronts were found to travel northward about 3 to 4 m per year and simultaneously to shift laterally. "According to changing predominant winds during the year, the crest is oscillating between an eastern and a western position." These results seem to confirm the sand movements expected with conflicting wind directions of the type recorded.

The wind systems in various parts of the Namib are clearly different, or at least represent different combinations and intensities of the main regimes. In the south, as at Luderitz for example, the winds are recorded as much stronger than at Gobabeb (Goudie, 1972, p. 16), and along the Atlantic the southwesterlies apparently are far more effective in transporting sand than they are inland, as demonstrated by the size and orientation of large transverse dunes in a coastal belt. Along the Kuiseb River in the north, the more nearly equal effects of three directions of wind presumably account for the many star dunes in that section of the sand sea.

In considering the wind system of the Namib Desert, the present wind regime should not be assumed to be necessarily or even likely the same as that of various times in the past. That winds are now much weaker than when the large linear dune pattern was formed was contended by Besler (1976, p. 84) on the basis that she did not consider winds of today capable of establishing this pattern. A last glacial maximum (about 19,000 to 13,000 years B.P.) was believed by Heine (1980, written commun.) to have been "extremely windy all over southern Africa. . . ." Furthermore, he stated that sand movement of the Namib erg during Pleistocene time was very intensive, but "during the Holocene only minor movements of the sand occurred."

Vegetation

The Namib sand sea, in common with most sand deserts that have scant rainfall, shows little vegetation much of the year except along erosion channels and dry river beds such as the Kuiseb. However, in the infrequent periods of concentrated rainfall, which come at highly irregular intervals over long periods of time, an amazing growth of vegetation is established extremely rapidly and greatly changes the appearance of the landscape. At the time of the present study the region had an extensive cover of plant growth.

Many of the interdunes in the interior are covered with grasses during periods of precipitation (Fig. 2*B*) or contain widely scattered shrubs (Figs. 2*C*, 2*D*). Because of high water tables, trees such as *Acacia*, *Euclea*, and *Salvadora* are common along major stream beds (Scholz, 1972, p. 34). In the fog belt adjacent to the Atlantic Coast and for some kilometres inland, surfaces tend to be salty because of the dew, and many of the plants that grow farther inland are absent. In general, vegetation is considerably more abundant among the interior red dunes than near the coast (Logan, 1960, p. 137), and these plants contribute toward stabilization of the dunes.

Geology

The geologic history of the Namib platform that underlies the desert has been discussed by many geologists (Logan, 1960; Scholz, 1972; Goudie, 1972; and others). The platform is composed principally of Precambrian rocks referred to the Damara System and consists of mica schist, gneiss, quartzite, and marble that have been dated as between 500 and 550 million years old (Clifford, 1967). These rocks are intensely folded and have been intruded by Precambrian igneous rocks known as the Salem Granite (Scholz, 1972, p. 34). This granite forms the small but prominent hills in the vicinity of Gobabeb.

Basic igneous rocks were extensively developed during Triassic and Jurassic times in parts of the region. These rocks are represented by basaltic dikes that today form linear ridges on the Namib platform and are considered part of the Karroo System (Scholz, 1972, p. 34).

The Namib platform, which was formed by extensive planation of the Precambrian and later rocks, has been referred to by various geologists as having "extraordinary flatness" and as being a "monotonous plane" (Logan, 1960, p. 89). Actually, this flat surface is an inclined plane that rises eastward from sea level to between 900 and 1,200 m at the Great Escarpment, broken only by scattered monadnocks or massive inselbergs that rise well above the plane. The retreating escarpment forms the eastern margin, with several large incised river valleys transecting the surface from east to west. Evidence of several lesser surfaces below the present upper one suggests a complex history of uplift and denudation, as discussed by Spreitzer (1965).

Cenozoic History

Principal events in development of the morphology of the Namib Desert consist of (1) forming of the flat surface of the Namib platform, (2) introduction of vast accumulations of sand on this surface, (3) re-elevation of the region with consequent incision of narrow stream valleys, and (4) development of large linear dune ridges with a general northward movement of the sand. Although most geologists are in agreement with the occurrence of these events, there seems to be a considerable difference of opinion concerning details of their development and the age or time of each.

Bevelling or planation of the ancient bed rocks to form

the remarkably flat and even surface of the Namib platform was referred to as a peneplanation by Logan (1960), King (1967), and others. Later writers described it as pediplane development (Selby, 1976, p. 6; Goudie, 1972). The time when this planation took place has generally been referred to the Late Cretaceous (Kaiser, 1926; Logan, 1960; King, 1967), although some geologists have considered it more recent, that is, early Tertiary (Selby, 1976, p. 6) or Pliocene and Miocene (King, 1967).

The time of re-elevation of the platform accompanied by stream incision was placed in the Miocene by King (1967), but in the early Pleistocene by Goudie (1972). The time of incision of the Kuiseb River and its tributaries was considered by Rust and Wieneke (1974, p. 11) to represent the first stage in a sequence of at least 12 morphoclimatic stages that they recognized in the Homeb area on the basis of land forms, fine sediments, and gravels.

The times of principal sand accumulation and the northward migration of sand bodies seem bo be matters of considerable conjecture, as few objective data for dating are available. A sand sea is believed to have existed in Pliocene time, and the Tertiary dunes were reworked into recent active sands in Pleistocene time, according to Barnard (1973, p. 10). Sand sheets accumulated in the Pliocene, developed into "active longitudinal dunes that moved northward in Pleistocene time" (Selby, 1976). Other suggestions concerning time of sand-sea development generally focus on Pleistocene time.

A Pleistocene development of the Namib sand sea must be related to climatic conditions during that epoch. Compelling evidence based on ecological studies by Koch (1961) suggests an extremely long duration of the present desert environment. "Superb endemism and adaptation of fauna" indicate undisturbed conditions of habitat for the highly specialized desert biota. In contrast, a moister climate during Pleistocene time was suggested by Goudie (1972, p. 26) on the basis of the types of faunas and sediments on raised beaches and of soil studies near Gobabeb by Scholz (1968).

A few dates have been obtained to elucidate some features of the relatively late history in the Pleistocene. A date greater than 60,000 years B.P. for a wide shallow lake at Narabeb, with indications of early man occupancy, has been suggested (Seely and Sandelowsky, 1974, p. 61). At a date of 10,000 years B.P., which is commonly considered the end of the Pleistocene Epoch, ponding is recorded on the basis of fresh-water snails from near Tsondab Vlei.

Sand Deposits

Sources of Sand

The sand of the Namib Desert probably had more than one source; evidence for at least three possible sources has been published. First, a major proportion of the desert sand probably has been introduced from central Africa by way of the Orange River to the south, as first described by Logan (1960, p. 133). Second, in the northwestern part of the Namib sand sea, a relatively small but distinctive contribution of sand has been accumulated as a mixture of deposits from the infrequent, violent floods of the Kuiseb River and the beach sands of the coast, as demonstrated by Nagtegaal (1973, p. 12). Third, a large source of red sand along the south side of the Kuiseb River is indicated by the bevelled tops of friable, red, cross-stratified sandstones cropping out within interdunes. These remnants occur at various localities such as southeast of Gobabeb and along the margins of Tsondab Vlei.

Concerning the southern sand source, "All of the interior of Africa east of the Great Escarpment drains southward to the Orange River . . . that carries a perennial flow westward through all the north-south trending ranges and fault blocks into the Atlantic Ocean. The interior is a land of torrential rains, sheet floods and gully erosion, and the Orange consequently carries a vast burden of sediment to the coast" (Logan, 1960, p. 133). With dominant winds from the south and southwest and with a system of large lnear dunes that have been migrating northward, sand from this major source clearly has been contributed for a long time.

The relatively small, crowded dunes of barchanoid and transverse types along the northwestern coast, formed largely by unidirectional winds from the South Atlantic, seem to be composed of a mixture of beach sand and Kuiseb River sand, as indicated by grain analyses made by Nagtegaal (1973), p. 11, 12). The angularity, large size of many grains, and composition of the sand led Nagtegaal to believe that the fluvial contributions were derived mostly from the Precambrian basement rocks exposed along the north side of the Kuiseb. In contrast, the bright red sands forming the giant linear and star dunes of the interior seem to display colors derived primarily from the red sandstones that crop out at numerous places within the dune field south of the Kuiseb (Fig. 46) and that are exposed east of Tsondab Vlei (Selby, 1976, p. 6; Besler, 1976, p. 84; Besler and Marker, 1979, p. 155).

The west-facing escarpment of friable red sandstone at Tsondab Vlei is composed of sand that closely matches in color and texture the bright red dunes of that area. It consists of horizontal and wavy beds, 0.3 to 1 m thick, that form a weak cliff about 60 m high; this cliff apparently has receded across a considerable area, furnishing a large supply of sand to the dunes farther west and northwest. Supporting this concept are records of the easterly berg winds that periodiccally blow with great strength off the plateau above. Isolated outcrops and residual hills of the red sandstone occur at numerous places along the vlei.

Principal Facies

Deposits of the Namib Desert that are composed partly or entirely of sand are classified as facies and represent the various environments in which each was formed. By far the most extensive of these facies is the great sand sea, consisting of dunes and interdunes; however, numerous other sand environments are associated with this facies. They include fluvial channels and flood plains, temporary lakes and ponds, tidal flats, lagoons or sebkhas, and beaches. Most of these facies border the dune field, but some are within it.

The dune-interdune facies has been variously subdivided according to its principal geomorphic forms into three or four units or belts, as described by earlier workers (Barnard, 1973, p. 10; Besler, 1976, p. 84). In general, the dune types are determined by the wind direction or directions in a particular area, but to some extent also by vegetation, outcrops, sand availability, or other features.

Individual dunes of transverse type attain heights of about 65 m in the coastal belt (Glennie, 1970, p. 94); linear and star dunes are 80 to 100 m near Gobabeb (Goudie, 1972, p. 24), and dunes mostly of star type are about 300 m near Sossus Vlei (Barnard, 1973, p. 2, 5). This height is controlled by the threshold velocity of the wind; above some undetermined height, star dunes that have vertical growth rather than forward growth commonly form on tops of large linear dunes. Small coastal dunes, formed by unidirectional winds, commonly migrate rapidly downwind, but the large linear dunes, believed to be controlled by winds of two directions, move in a longitudinal direction very slowly. Because of counter winds, these linear dunes commonly become reversing dunes with slipfaces that dip first one direction, then the opposite.

Sands deposited by rivers on flood plains (Scholz, 1972, p. 51) and in stream beds probably represent a large proportion of the sediment being introduced into the area. The Orange River on the south and the Kuiseb River on the north are the two major drainages that extend across the dune field to the Atlantic, and the Orange, in particular, transports a tremendous load of sediment from central southern Africa. Other rivers that probably once crossed the dune field now are blocked by the giant linear dunes, to form terminal playa-like deposits at their end points (Logan, 1960, p. 102). The intermittent stream courses resemble, in general, wadis of north Africa and washes of the southwestern United States.

Additional sand facies that developed in the interior parts of the sand sea are ephemeral ponds or small lakes, formed where water is lost by seepage and evaporation in some interdune valleys (Logan, 1960, p. 135). One very large temporary lake, with traces of early man along its original shoreline, apparently was caused by the former Tsondab River (Selby and others, 1979).

Sand facies of coastal areas adjacent to the dune fields consist of three principal types: tidal flats, lagoons, and beaches (Figs. 3*A*, 3*B*). Supratidal sand flats in the Walvis Bay region are described by Nagtegaal (1973, p. 7) as "an extensive, almost featureless plain composed of sand . . . estimated to be between 1 and 2 m above mean high tide." Beach sands and those of coastal lagoons that form along the windward side of the dunes have been examined in some detail by Nagtegaal (1973, p. 11) who discussed their mineral composition and texture.

Both alluvial and beach sands of the region, based on the few samples available (Table 1) showed dominant sizes ranging between fine and medium grains (⅛ to ½ mm); this size is similar to that of most of the dune sands of the sand sea.

Sand Texture

Grain size and shape seem to be relatively uniform throughout the Namib sand sea—at least in the northern part where this study was conducted. Of 46 selected samples representing an area extending from the Atlantic Ocean eastward to the Great Escarpment, approximately half, measured on the Wentworth scale, are dominantly medium grained (¼ to ½ mm) and half are dominantly fine grained (⅛ to ¼ mm). Only small percentages of all other sizes are included. Results of these analyses are summarized in Table 1, arranged by dune types. A representation of interdunes is included. Sorting, according to the Payne scale, is good in all samples examined, but the interdunes, in general, show a much greater range in grade size than do the associated dunes.

A comparison of grain size for the various types of dunes suggests that the large linear dunes of the interior are composed largely of medium-grained sand, whereas the interdunes between them and transverse dunes that locally cross the interdunes consist of much greater proportions of fine sand. Coastal barchanoid dunes also contain a considerable percentage of fine sand. A comparison of dune crest sand from Gobabeb and coastal dune sand from Swakopmund was made by Goudie (1972, Fig. 10), who showed sand in the interior dunes as coarser than the coastal sand.

In a study of grain sizes in linear dunes of the Namib dune field, Lancaster (1981, p. 6, Fig. 5) noted from his analyses that sand of the crests and slipfaces was both finer and better sorted than that forming the plinths at their bases. This distribution pattern according to size grade Lancaster attributed to sand movement dominantly by saltation on the crests and flanks and by surface creep or traction on the plinths.

Detailed textural studies by Nagtegaal (1973, Fig. 3) in the northwest coastal area compared size distribution of beach, adhesion-ripple (interdune), and dune sands and showed the maximum grain size in beach sands and coastal dune sands to be similar. In adhesion ripples it is somewhat greater. Other studies in the same report compared the size of heavy minerals in Kuiseb River sands with those in dune, interdune, and beach facies, and these studies discussed the roundness of sand grains in these environments. Beach sands were reported to be predominantly subrounded but with a small amount of subangular grains. Both adhesion-ripple (interdune) sands and dune sands contain large amounts of angular grains (Nagtegaal, 1973, p. 11).

Mean values expressed in the percentage of grain size for dune sands of the Namib have been compared with those of the southern Kalahari Desert to the east by Goudie (1972, Table 3). A conclusion was reached that the sizes in both deserts were fairly similar, ranging between medium (¼ to ½ mm) and very fine (⅛ to ¼ mm) grained in about the same percentages.

Sand Composition

Sand of the Namib Desert is in most places composed of a very high percentage of quartz grains. In one area south of the Kuiseb River, at Soutrivier, however, two types of dune sand (light and dark) are described by Scholz (1972, p. 36) as composed mainly of quartz and feldspar and of garnet, monazite, feldspar, and opaque ore minerals, respectively. Secondary constituents included chlorite, zircon, epidote, staurolite, tourmaline, and ilmenite.

Near Walvis Bay at the mouth of the Kuiseb River, a study of heavy minerals to compare sands of the beach,

TABLE 1. GRAIN-SIZE DISTRIBUTION (WENTWORTH SCALE) IN VARIOUS SAND BODIES ASSOCIATED WITH NAMIB DESERT SAND SEA

Sample description: Locality	Position	Sample No.	Percentage of sand in one-half phi units: −0.75 Gravel	−0.75 Very coarse	−0.25	0.25 Coarse	0.75	1.25 Medium	1.75	2.25 Fine	2.75	3.25 Very fine	3.75	<3.75 Silt and clay
							Beach sand							
6 km S of Swakopmund	Upper foreshore	21					1.0	22.4	57.5	17.4	0.5	0.3	0.3	0.6
Do.	Backshore	22		0.3	1.1	2.1	2.8	24.0	49.3	15.6	3.0	0.9	0.3	0.6
Do.	500 m from berm (backshore)	23	25.5	5.7	0.8	0.6	1.3	7.9	12.3	19.2	15.2	7.1	2.8	1.6
							Alluvial sand							
Tsondab Vlei	Floodplain channel	51	0.6	0.4	1.3	0.9	0.6	3.0	9.1	16.2	31.9	22.7	8.4	4.9
Do.	Margin of floodplain	54			0.1	0.7	44.0	19.4	23.9	10.9	0.7	0.1	0.1	0.1
							Linear dunes							
41 km S of Nara Valley	W side, 26 m up	62				0.1	4.4	37.2	20.6	14.1	12.2	8.1	2.9	0.4
Do.	E side, 4 m up	63					1.5	70.7	20.4	4.5	2.0	0.6	0.1	0.2
23 km S of Nara Valley	W side, 31 m up	58				0.1	4.7	52.4	21.6	10.3	5.1	3.5	1.9	0.4
Do.	E side, 4 m up	57				0.1	6.5	32.5	46.2	11.7	2.2	0.5	0.1	0.2
6 km S of Rooibank	W side, upper	47						2.2	19.9	27.3	33.1	14.1	3.2	0.2
Do.	W side, middle	46				0.1	12.6	28.3	23.9	14.1	14.2	5.4	1.3	0.1
Do.	E side, upper	44				0.1	1.6	25.0	22.5	23.7	21.0	5.2	0.7	0.2
Do.	E side, middle	43				0.1	15.9	35.1	20.7	13.1	11.5	3.2	0.2	0.2
							Star dunes							
17 km NW of Gobabeb	Upper windward slope	33				0.1	0.5	16.8	48.4	23.3	8.2	2.2	0.3	0.2
Do.	Top of slipface	34						0.2	55.6	38.0	3.8	1.7	0.4	0.3
Tsondab Vlei	Base of slipface	55				0.1	4.3	40.5	23.3	16.6	11.1	3.8	0.1	0.2
Do.	Base of slipface	52					0.1	11.9	61.1	22.6	2.9	0.9	0.3	0.2
							Barchanoid dunes							
10 km E of Walvis Bay	Lower slipface	28					0.5	14.9	61.8	21.5	0.4	0.2	0.3	0.4
Do.	Base windward slope	30		0.1	0.2	0.1	0.2	4.6	39.6	43.9	4.6	3.3	2.0	1.4
Do.	Upper windward slope	31				0.1	0.3	6.2	32.9	55.8	3.4	0.2	0.4	0.7
Do.	Crest	32			0.1	0.2	0.6	11.8	59.4	25.7	1.5	0.1	0.2	0.4
5 km N of Walvis Bay	Lower slipface	39		0.1	0.1	0.2	1.7	16.1	23.7	21.0	23.1	11.8	2.0	0.2
10 km N of Walvis Bay	Lower windward slope	36	0.7	6.4	2.1	0.1	0.1	0.2	43.9	34.3	3.5	3.8	3.3	1.6
6 km S of Swakopmund	Middle slipface	24	0.4	0.2	0.2	0.3	9.3	58.0	21.8	7.0	1.7	0.3	0.3	0.5
							Transverse dunes							
Nara Valley	3 m below crest, windward side	17				0.2	0.2	3.5	11.4	44.9	31.5	7.3	0.6	0.4
Do.	Near top, lee side	15							3.2	44.9	31.5	7.3	0.6	0.4
Do.	3 m down, lee side	38				0.1	0.1	2.6	23.8	46.2	20.3	5.8	0.9	0.2

TABLE 1. GRAIN-SIZE DISTRIBUTION (WENTWORTH SCALE) IN VARIOUS SAND BODIES ASSOCIATED WITH NAMIB DESERT SAND SEA

Sample description		Sample	Percentage of sand in one-half phi units											
			-0.75	-0.75	-0.25	0.25	0.75	1.25	1.75	2.25	2.75	3.25	3.75	<3.75
Locality	Position	No.	Gravel	Very coarse		Coarse		Medium		Fine		Very fine		Silt and clay
						Dome-type dune								
2 km S of Gobabeb	Lag, dune base	61		5.6	8.9	11.4	23.1	29.0	10.9	5.5	3.2	1.3	0.7	0.4
Do.	0.5 m below crest windward side	27	0.2	0.1	0.0	0.1	0.2	23.9	42.6	5.5	7.7	6.0	3.7	0.9
						Blowout dune								
Tsondab Vlei	Inside crater, NE side	53	16.2	3.4	0.2	0.6	2.3	8.2	9.5	18.0	23.1	11.1	3.9	3.5
						Interdune								
S of Walvis Bay	Mud-sand sebkha	25		1.2	1.2	1.2	2.5	6.9	12.1	19.3	24.2	14.6	4.9	4.5
10 km N of Walvis Bay	E side, coastal dunes	35	40.0	5.6	3.3	1.0	0.8	1.2	9.1	10.5	4.7	9.9	9.6	4.3
10 km E of Walvis Bay	Between barchans	29	29.5	7.4	6.1	5.9	5.0	6.1	3.3	2.6	16.7	12.4	4.4	0.6
6 km S of Rooibank	Center of valley	45	0.2	0.7	0.0	0.4	4.5	18.7	26.7	27.4	8.6	1.7	0.1	0.1
SW of Rooibank	Valley margin	48	0.4	0.2	0.2	0.2	1.4	7.4	24.7	38.7	18.7	5.2	1.6	1.3
Narabeb	E margin	42	18.8	0.6	0.9	1.4	8.5	16.1	20.3	20.7	9.3	2.0	0.3	1.1
Do.	Center of valley	59	2.7	0.2	0.4	2.0	30.2	22.3	18.2	13.4	5.1	3.1	1.3	1.1
23 km S of Nara Valley	do.	37		0.1	0.4	2.3	16.8	11.8	22.3	16.6	11.7	10.5	5.7	1.8
Narabeb	W part of valley	40		0.1	0.8	8.1	34.4	22.5	8.0	5.7	7.6	8.3	3.7	0.8
Do.	E part of valley	41	15.0	0.7	1.0	3.4	10.1	13.1	12.8	15.9	11.2	8.4	4.5	3.9
11 km S of Nara Valley	East margin	14	15.3	0.2	0.2	0.4	3.9	18.5	9.8	13.4	21.0	12.1	3.8	1.4
Do.	Near dune base	16		0.2	8.0	25.3	25.8	18.1	5.4	5.4	6.4	4.1	0.9	0.4
Soutrivier	E side of valley	56		0.1	0.1	0.2	2.8	24.7	43.4	21.5	4.6	1.1	0.5	1.0
5 km W of Gobabeb	N end, near Kuiseb	20	4.6	3.3	0.2	0.2	5.6	21.2	8.7	25.9	21.5	6.7	1.3	0.8
2 km SW of Gobabeb	E part of valley	26	15.3	1.7	0.4	0.2	1.1	1.4	8.7	18.8	26.8	15.0	6.5	4.1
10 km SW of Gobabeb	W part of valley	18	19.1	0.1	0.2	0.3	3.6	7.4	4.9	8.8	14.2	16.6	14.3	10.5
Do.	Center of valley	19	19.0	0.1	0.1	0.3	2.8	9.9	7.1	10.9	19.4	13.8	8.8	7.8
5 km S of Homeb	NE of dune base	50					7.9	63.6	17.5	7.6	2.9	0.3	0.1	0.1
Do.	Center of valley	49		0.8	0.5	0.2	2.7	9.9	15.0	18.3	21.1	12.9	7.6	6.6

coastal dunes, adhesion ripples (interdunes), and river bed was made by Nagtegaal (1973, p. 12). Sillimanite was found to be the major heavy-mineral component in all four sand facies, but garnets and hornblende also were present in various amounts. Other minerals recorded were zircon, rutile, zoisite, and cordierite.

Ripple Marks

Eolian ripple marks are a common feature throughout the Namib sand sea, as described by Logan (1960, p. 136), and many varieties in size and shape are represented. Wind ripples are distinctive both because of their shape, recorded in terms of height to width (ripple index), and their orientation with respect to the principal dune surfaces (McKee, 1945, p. 317–320). They range in size from normal types a few centimetres high (Figs. 4*A*, 4*C*) to megaripples (granule ridges) that are several times larger and that contain relatively coarse grains (Fig. 4*B*). Mostly they form patterns of parallel, nearly straight to gently undulating or wavy ridges that cover broad surfaces.

Despite their abundance on dune surfaces, the proportion of rippled deposits preserved within a dune probably is relatively small. Most ripples formed by a unidirectional wind on the gentle slop of the windward side are ephemeral, for such deposits are constantly being destroyed as the dune migrates. On the lee side or sides of dunes, where ripple marks form on the steep slopes, at right angles to the dune crest, chances of preservation may be good. Where cohesion develops from moisture and rapid burial takes place from grainfall or from avalanching, ripple marks can become a part of the record.

A

B

Figure 3. Sand facies along coast at Sandwich Harbor, South West Africa. *A*, dune foreground, lagoons, tidal flats, and Atlantic Ocean in distance; *B*, tidal sand in foreground, windward face of transverse dunes in background.

Variability of wind strength and direction in many parts of the Namib is illustrated by ripple patterns, in addition to the multidirectional dune types. Normal ripple ridges, at right angles to megaripple crests and formed in their troughs (Fig. 4*B*), are not uncommon. Straight, parallel ripple crests, oriented at right angles to dune ridges, on steep foresets (Figs. 4*C*, 4*D*) are recorded from both linear and star dunes.

Adhesion Structures

In coastal parts of the Namib sand sea, especially in the interdune sebkhas of the northwest margin, various types of adhesion structures are abundant. One type consists of irregular corrugations, referred to as anti-ripplets by van Straaten (1953) and subsequently named adhesion ripples by Reineck (1955). A second type consists of thin layers recognized by Hunter (1980) and later referred to as adhesion laminae by Kocurek (1981). A third type that consists largely of irregular bumps has been termed adhesion warts by Reineck (1955), and it probably is the commonest form in the Namib (Fig. 5). All these structures develop primarily by dry, blown sand adhering to wet surfaces of one type or another, and thus accumulating to form either a lenticular mounded structure or an irregular drape covering small-scale topographic features.

Because adhesion structures are relatively distinctive and easy to recognize in a stratigraphic sequence, they seem to be good indicators of interdune environments like those in coastal areas of the Namib. In the Permian Rotliegendes of northwestern Europe, probable adhesion structures have been recognized and described by Glennie (1970, 1972).

Slump Structures

The term "slump structure" is used here in referring to sand masses that have avalanched down a dune foreset or slipface, at or near the angle of repose, with enough cohesion to retain a compact character. Two principal varieties of slumping are commonly recognized. In one, the sand mass breaks loose at its upper end, leaving a narrow trough, and forms a tongue at the lower end that overrides an older surface. In the second type of slumping, apparently the result of cohesion extending to greater depth, a series of narrow cracks parallel to the dune crest develops. Tension in the upper foreset slope causes the sand sheet to pull apart into segments and to develop a steplike form.

In interior parts of the Namib sand sea, much of the avalanching on dune slipfaces is by grain flowage because of the extreme dryness of sand most of the time. In areas near the coast and as far inland as fog or dew is common, slumping is normal. Most slump marks are of the tonguelike variety (Figs. 6*B*, 6*D*), although the parallel pull-apart type forms locally.

Slump surfaces of the tonguelike variety are especially abundant on the steep side of star dune arms south of Homeb, where they bury both smooth and ripple-marked surfaces. Also, in places they overlap one another to form long parallel linear bands from the top nearly to the bottom of a slope (Fig. 6*A*). Dune crests where the slumps begin form scallops as individual avalanches pull loose. Beetle tracks crisscross the slump surfaces (Fig. 6*C*), and in places the tracks even go directly up the steep 34° slopes.

As a result of large-scale slumping, various types of internal deformational structures are formed within the deposits, as seen in trenches. These structures include faulting, folding, brecciation, and contortion (Figs. 7*A*, 7*B*), as described in the following sections.

Deformational Structures

Few deformational structures were observed in the dry dunes of the interior Namib, but such structures were fairly numerous in coastal areas where heavy fogs are common and dunes frequently are wet. Various types of structures are represented, but most or all of them probably are related to the avalanching, especially slumping, of slipfaces (McKee and others, 1971). Overloading with sediment from above may have been responsible for some.

In linear dunes south of Rooibank, in strata dipping at 29° and 30°, excellent illustrations of stairstep (monoclinal) folds (Fig. 7*A*) occur, presumably as a result of slumping. On the lee side of a large transverse dune at Sandwich Harbor, examples of overturned folds, strata dragged during thrusting, and contorted bedding, possibly the result of loading (Fig. 7*B*) were observed.

Trace Fossils—Termite(?) Tubes

Long cylindrical tubes of agglutinated sand (Fig. 8), believed to have been formed by dune termites, were found in abundance in the interdune at Soutrivier and on the windward slopes of dunes 7 km south of Gobabeb and in Narabeb Valley. They also occur in many other places. Most of the tubes ranged between 5 and 7 mm in diamter, some were straight with lengths of 20 cm or more, others were curved, and some were branching (Fig. 9). A few bulbous forms were found and many had hollow centers. Structures of similar type, referred to as cylindrical shaped concretions, were recorded by Scholz (1972, Fig. 22).

In interdune areas, the tubes are mostly localized around grass hummocks (*Stipagrostis sabulicola*) where, as interpreted by F. von Blottnitz (1977, written commun.), grass has been encased with a layer of secreted matter and sand. Later, the termite ate out the grass leaving a hollow center. Large chambers of bulbous type presumably enclosed the bases of grass clumps.

On dune flanks as much as 20 m from the nearest live

A

B

Figure 4 (this and facing page). Eolian ripple marks in dunes of Namib Desert. *A*, parallel wavy type in magnetite sand on windward slope of barchan, north of Walvis Bay; *B*, megaripples, with regular ripples in troughs; interdune at Tsondab Vlei; *C*, crest on arm of star dune, corrugated ridge marking top of ripple series on present windward side; ripple length, crest to crest, about 3 cm; *D*, crest of linear dune with ripple series on both sides indicating two slipfaces affected by cross winds. Photographs by E. Tad Nichols.

C

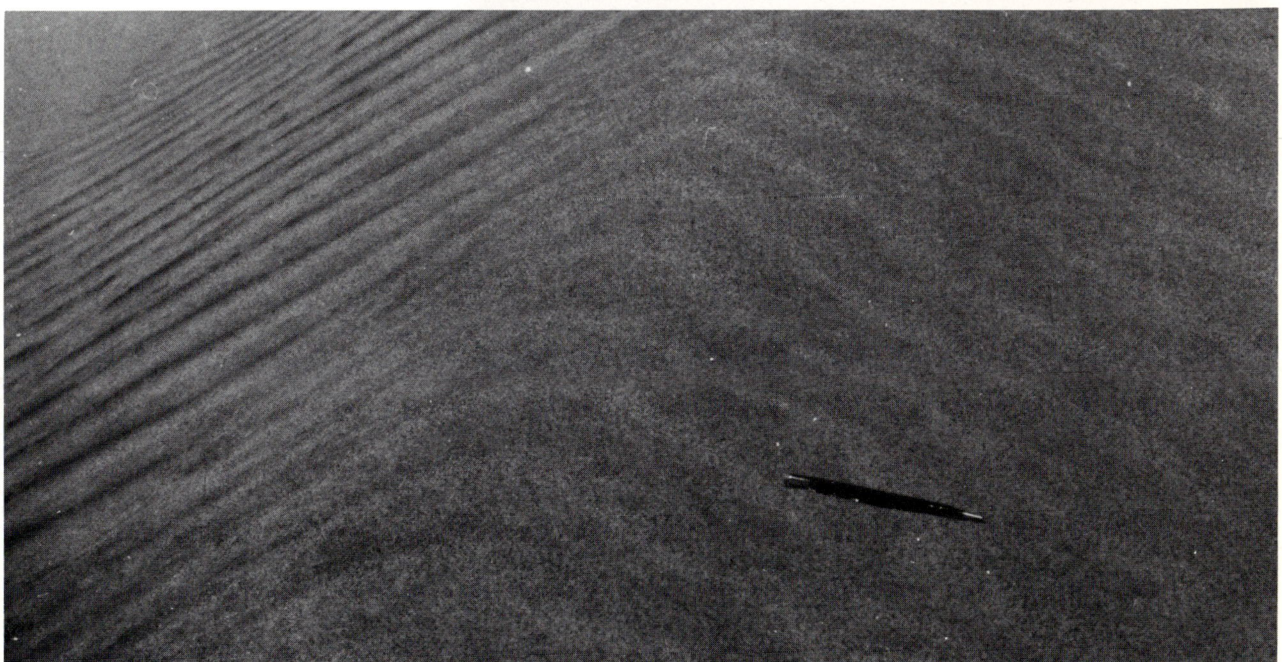

D

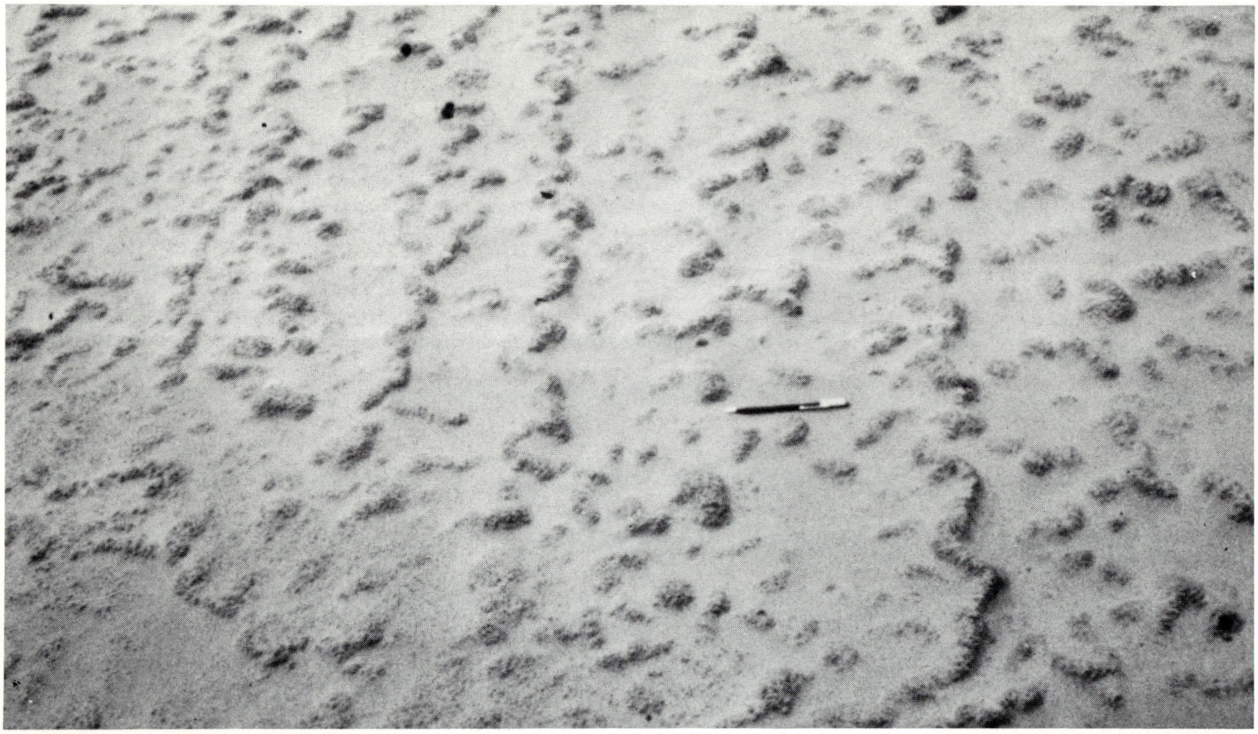

Figure 5. Adhesion structures on floor of interdune consisting of sebkha deposits between barchans, 3 km south of Walvis Bay.

grass of the interdune, many tubes were found *in situ* by trenching. One of these tubes, with a sinuous form (Fig. 9*B*), was traced for 2 m beneath a ripple-marked surface that dipped 14°. Some tubes were as much as 13 cm below the surface; none showed any appreciable tapering as in root casts. A few of those southwest of Gobabeb were nearly halfway up the windward slope of a dune.

The growth of minute calcite crystals (Fig. 10) coating the quartz grains inside some tubes indicates that enough moisture was present and sufficient pore space available to permit calcite growth.

Trace Fossils—Tracks and Trails

Large numbers of tracks and trails formed by insects, especially beetles, by reptiles, and by some larger animals (Scholz, 1972, p. 36) are found in the dry sand of dune slipfaces in many parts of the Namib Desert. Such impressions commonly are sharp and clear and are the potential record of a dune fauna. They may lead laterally across a 28° to 30° slope, or they may go nearly straight up the steep slope (Fig. 11). For permanent preservation, the bedding surface on which they are formed probably requires some degree of cohesion at the time the tracks are buried by additional sand layers (Ahlbrandt and others, 1978). Such cohesion seems to be furnished by the almost daily heavy fogs that come in from the South Atlantic to wet the dune surfaces. In this respect, the Namib probably is one of the most favorable sand seas for both the habitation by animals and the preservation of their traces.

Beetles which are among the most common of the animals to leave tracks on the sand come out on the dune surfaces during the day, but especially at dusk or later, and accumulate quantities of organic material blown in by wind (Koch, 1961). This also serves as a record of dune life that may be preserved.

Dune Color

The color of sand in the Namib Desert is one of its most striking features and has been commented on by virtually every geologist to work there. A marked contrast between the light or yellowish brown of the coastal areas and the bright red brown of the interior has been the subject of much discussion. The coastal dunes have variously been described as light yellowish brown (Barnard, 1973), light brownish color (Scholz, 1972), and yellow-white (Logan, 1960). In contrast, the interior dunes are referred to as red by Barnard, reddish brown by Scholz, and brick red by Logan.

An explanation suggested by Logan (1960) for the variation in sand coloration involved differences in age. According to this hypothesis, dunes of the interior that he considered to be older than those of the coast supposedly had

A

B

Figure 6 (this and following page). Slump marks on dune slipfaces in Namib sand sea. *A*, parallel-band type, 1 km south of Homeb; *B*, tonguelike form, 6 km south of Kuiseb, Tsondab Vlei; *C*, beetle tracks on recent slumps; tracks give scale; Tsondab Vlei; *D*, tonguelike form, 1 km south of Homeb. Photographs by E. Tad Nichols.

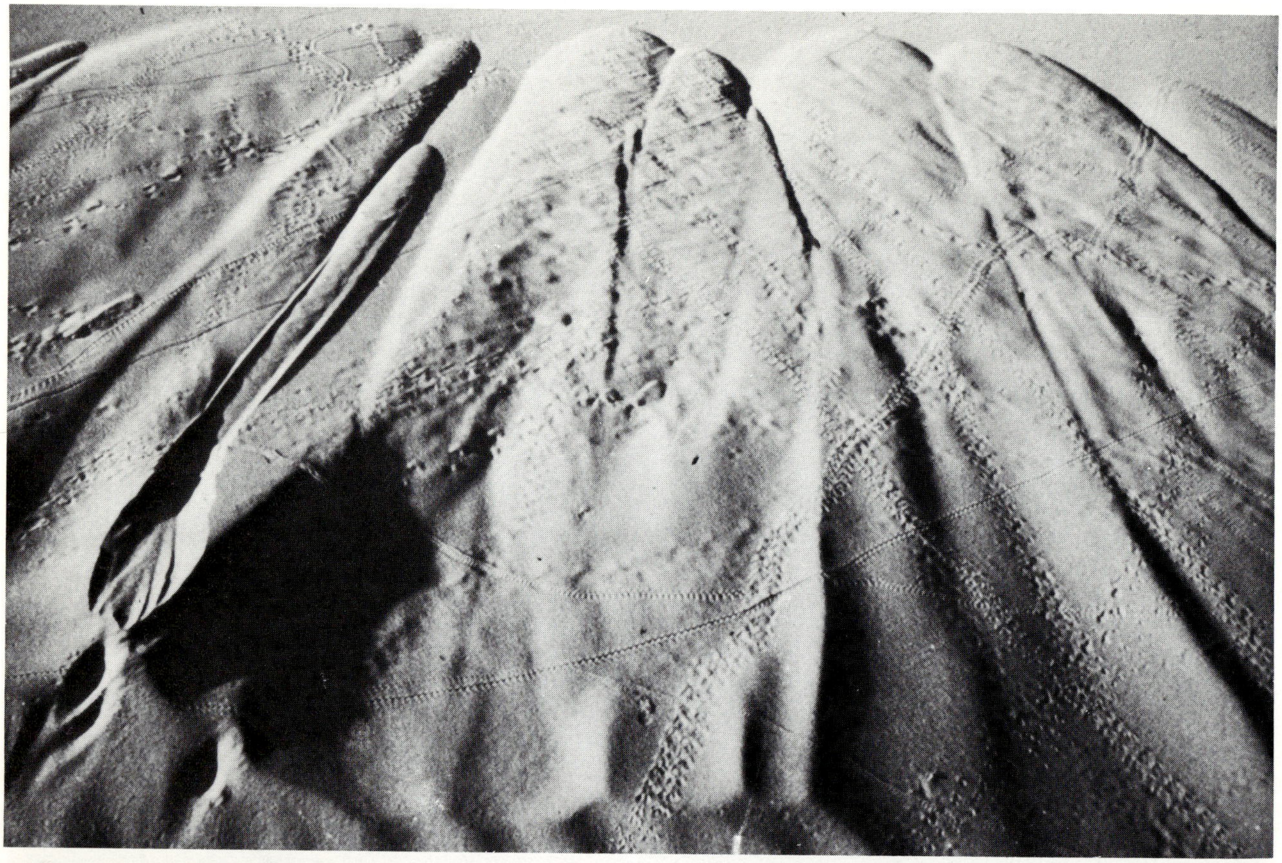

C

D

A

B

Figure 7. Deformational structures on dune foresets, Namib sand sea. *A*, stairstep (monoclinal) folds, south of Rooibank; *B*, contorted bedding, possibly from overloading, transverse dune at Sandwich Harbor; scale in inches. Photographs be E. Tad Nichols.

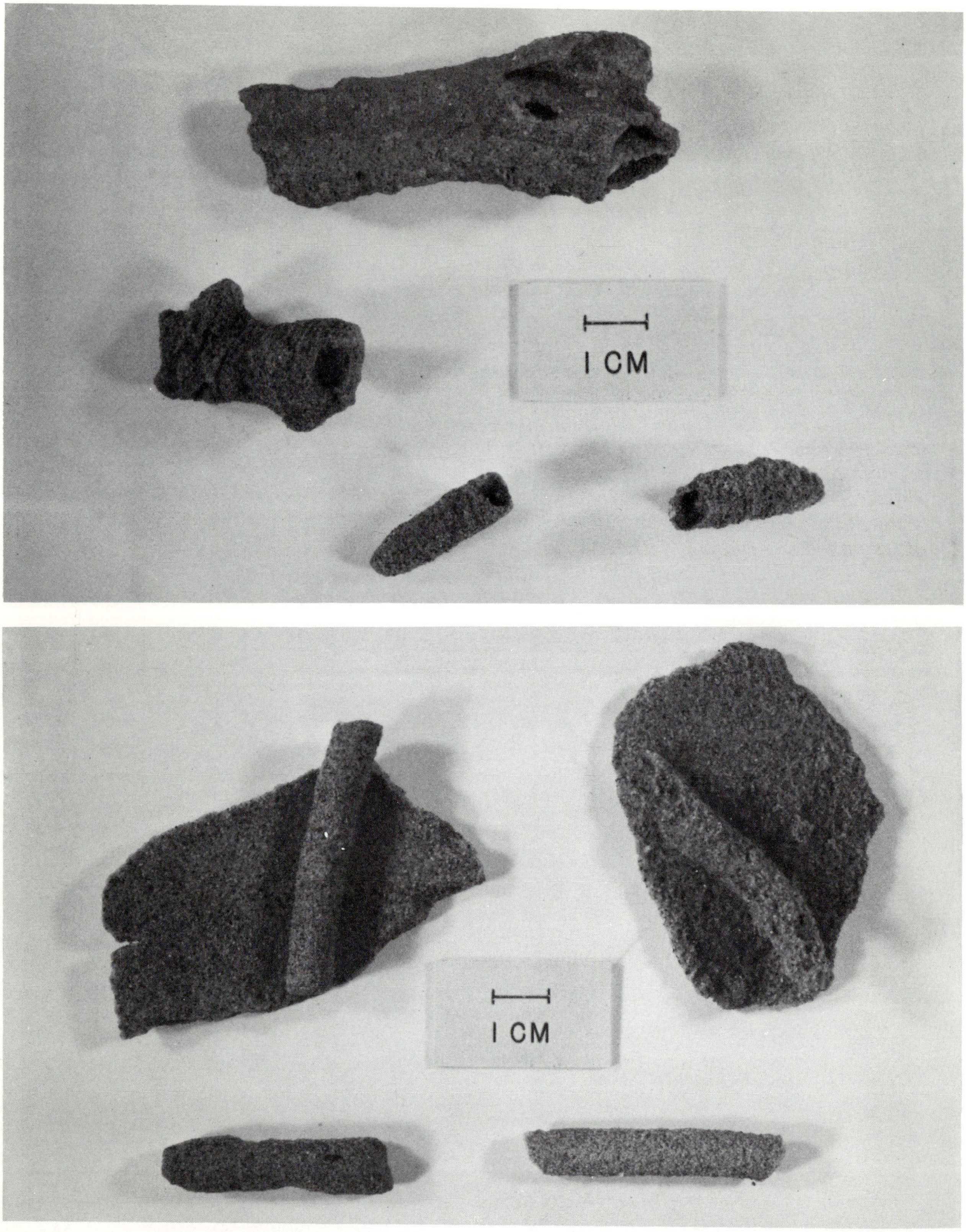

A

B

Figure 8. Termite(?) tubes from beneath surface of dry, ripple-marked sand on windward slope of linear dune, 7 km southwest of Gobabeb. *A*, tubes showing hollow centers and agglutinated sand sheaths; *B*, straight to gently curving sections of cylinders forming termite tubes.

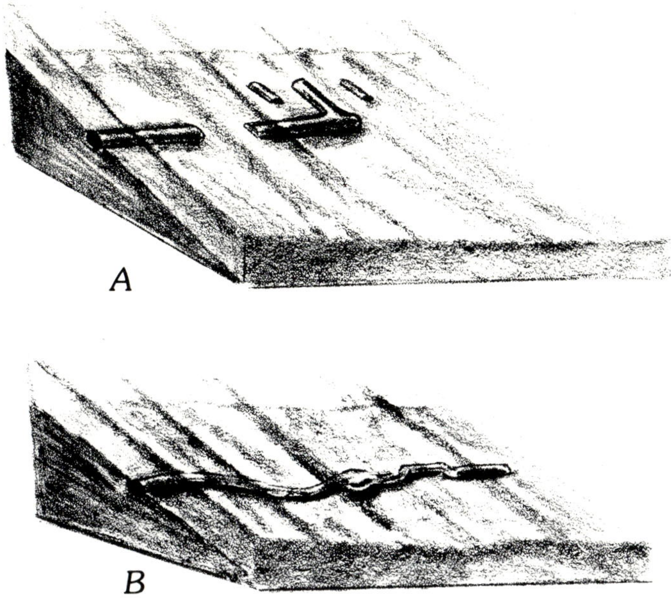

Figure 9. Termite(?) tubes (natural size) from flank of linear dune, 7 km southwest of Gobabeb. *A*, branching form *in situ* on 14°-dip slope of rippled sand; *B*, irregular, undulating form about 2 m long and 1 cm in diameter, attitude nearly horizontal, 13 cm below surface. Sketch by T. Kostick.

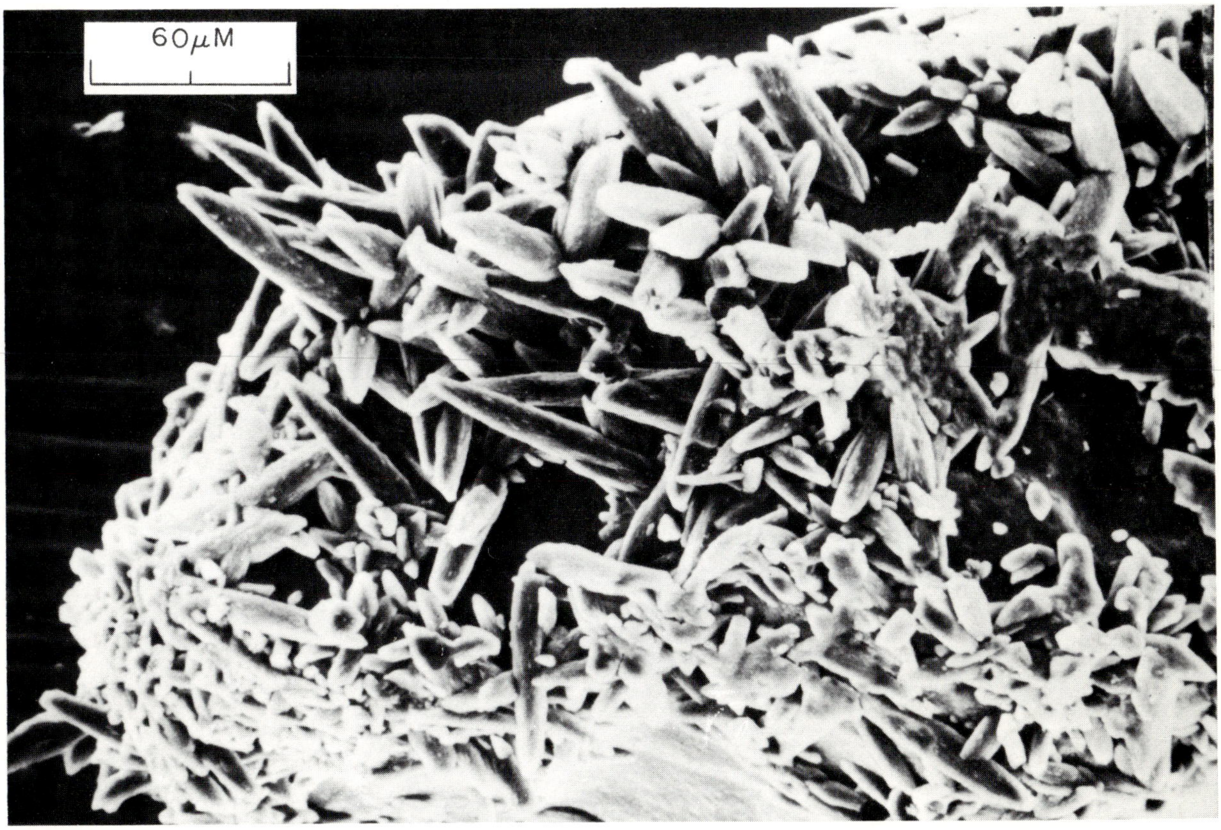

Figure 10. Scanning electron photomicrograph showing calcite crystals coating sand grains in termite tube, 7 km southwest of Gobabeb.

developed greater oxidation of iron on grain coatings because of greater age. However, time is only one of many factors controlling the red coloration of desert sands, as has been demonstrated by Walker (1979). Furthermore, as pointed out by Besler (1976, p. 88), sands of the coastal dunes have had a different origin from those of the interior (Fig. 46). Deposition in each place was controlled by a distinct wind system and involved sands that represent different sources as shown by grain roundness and mineralogy (Nagtegaal, 1973, p. 12). The red Namib Sandstone (Besler and Marker, 1979), discussed later under Red Hills in Interdunes, was a source of much bright-red dune sand in interior areas.

Although dune colors seem to change gradually from east to west across the region—probably the result of lateral gradation between sand from two sources—there also has been noted some intertonguing between light- and dark-colored sands in dune ridges west of Gobabeb. Furthermore, even within certain individual dunes, differences in color resulting from concentrations of certain dominant minerals are reported. Dune sides streaked with chocolate brown from minute grains of ilmenite and magnetite were recorded by Logan (1960, p. 136). Different colors on an individual dune were stated by Scholz (1972, p. 36) to occur with red concentrations from garnet and dark brown from magnetite.

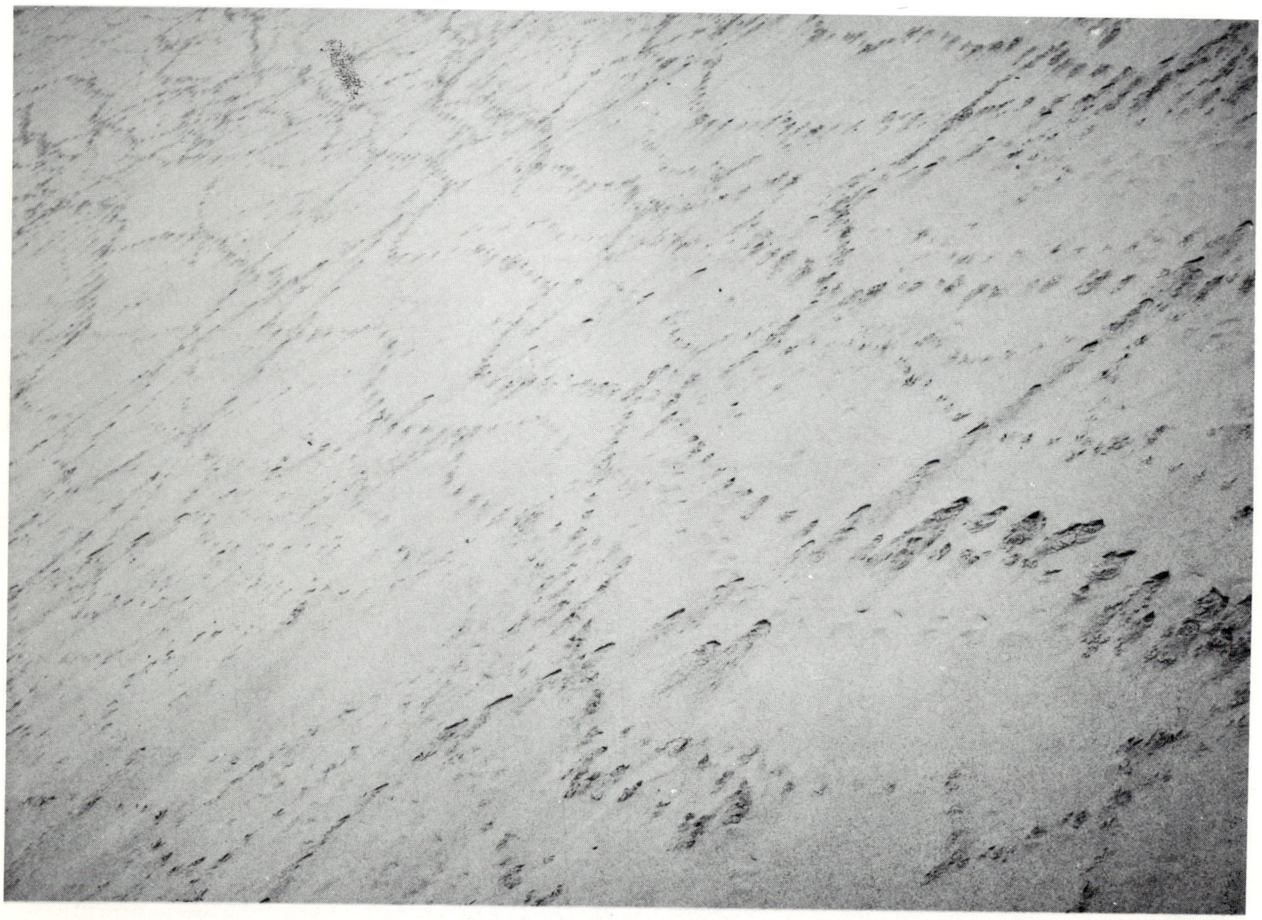

Figure 11. Bird tracks on dune slipface; barchans near coast, 3 km south of Walvis Bay.

Dune Types

Most of the major dune types that have been recognized throughout the world are represented in the sand sea of the Namib Desert. Differences in dune types probably result mostly from variations from place to place in wind regime. Whether the wind in any particular part of the desert is unidirectional, bidirectional or multidirectional seems to be the principal control of dune form, although physiographic features, vegetation or lack of vegetation, and amount of sand available may be locally important.

The classification of dune types as described and illustrated by McKee (1979, p. 8–13) is used in the following descriptions of dunes (Table 2). Thus, dunes formed by unidirectional winds include domes, barchans, barchanoid ridges, transverse ridges, and parabolic dunes, and they characteristically develop a single slipface. Dunes believed to result from bidirectional winds consist of a linear type (referred to in some regions as seif or as longitudinal) and a reversing type. Both types have two slipfaces. Dunes formed from multidirectional winds have three or more slipfaces and many variations in form. They are commonly called star dunes.

In the Namib Desert, the dominant dune type and by far the most widely distributed is the linear. It occupies most of the area from Luderitz in the south to the Kuiseb River in the north. Other dune types occur mainly along the margins of the desert where their development seems to be controlled by such features as the Atlantic Ocean to the west and the Great Escarpment of the interior plateau to the east. Parabolic dunes that are generally caused by stabilization through vegetation are reported by John Ward (1980, written commun.) from the coastal area along the western edge of the dune field, but have not been observed by me.

TABLE 2. TERMINOLOGY OF BASIC (SIMPLE) DUNE TYPES AND OTHER EOLIAN DEPOSITS (COMPOUND AND COMPOSITE FORMS NOT INCLUDED HERE)

Geometry	Number of slipfaces	Name based on geologic structure
Circular or elliptical mound	0*	Dome
Broad flat surface	0	Sheet
Elongate strip	0	Stringer
Crescent mound	1	Barchan
Row of connected crescents	1	Barchanoid ridge
Asymmetrical ridge	1	Transverse ridge
Circular rim of depression	1 or more	Blow out†
"U"-shape in plane view	1 or more	Parabolic†
Symmetrical ridge	2	Linear (Seif) longitudinal
Asymmetrical ridge	2	Reversing
Central peak with 3 or more arms	Several	Star

Notes: Modified from McKee (1979).
Definition of dune classes: Simple, dunes in the forms shown above; varieties, dunes having differences in details of topographic form; compound, dunes in which two or more of the same type are superimposed; complex, dunes having two or more different types combined.

*Internal structures may show embryo barchan-type with one slipface.
†Dunes controlled by vegetation in many places.

Attempts to subdivide the Namib sand sea into zones or belts of various dune types have been made by several geologists. Three divisions, recognized by Barnard (1973, p. 10) consist of (1) coastal transverse and barchan dunes, (2) interior linear dunes, and (3) eastern marginal dunes, mostly of star and compound types. These dune belts, extending in a north-south direction, were also noted by Besler (1976), p. 84).

Method of Study

This investigation of dune structures was made primarily by using a direct approach, that is, by exposing to view the strata and cross-strata of various selected dunes and interdunes through trenching. These internal structures, when considered in three dimensions, probably constitute the most important basis for distinguishing the principal types of dunes (McKee, 1957).

In the trenching program, various sand bodies to be examined were made cohesive by pouring water onto the surface of the desired sample site, after which a section was cut through the sand with a shovel (Figs. 31*B*, 35, 38). Most trenches were made with vertical sides and flat floors to expose structures both parallel to and normal to the dune crests. They were made either L-shaped or T-shaped in order to show stratification in three dimensions. The patterns of the structures were then recorded by measuring and sketching, by photography or by preparing latex peels. By locating trenches on two or more sides of a dune, on its crest, and at various elevations above its base, the general character of a dune could be ascertained by projection, and the direction or directions of migration could be determined.

Linear Dunes

Linear (longitudinal) dunes are not only the dominant and most widespread dune type in the Namib sand sea, but

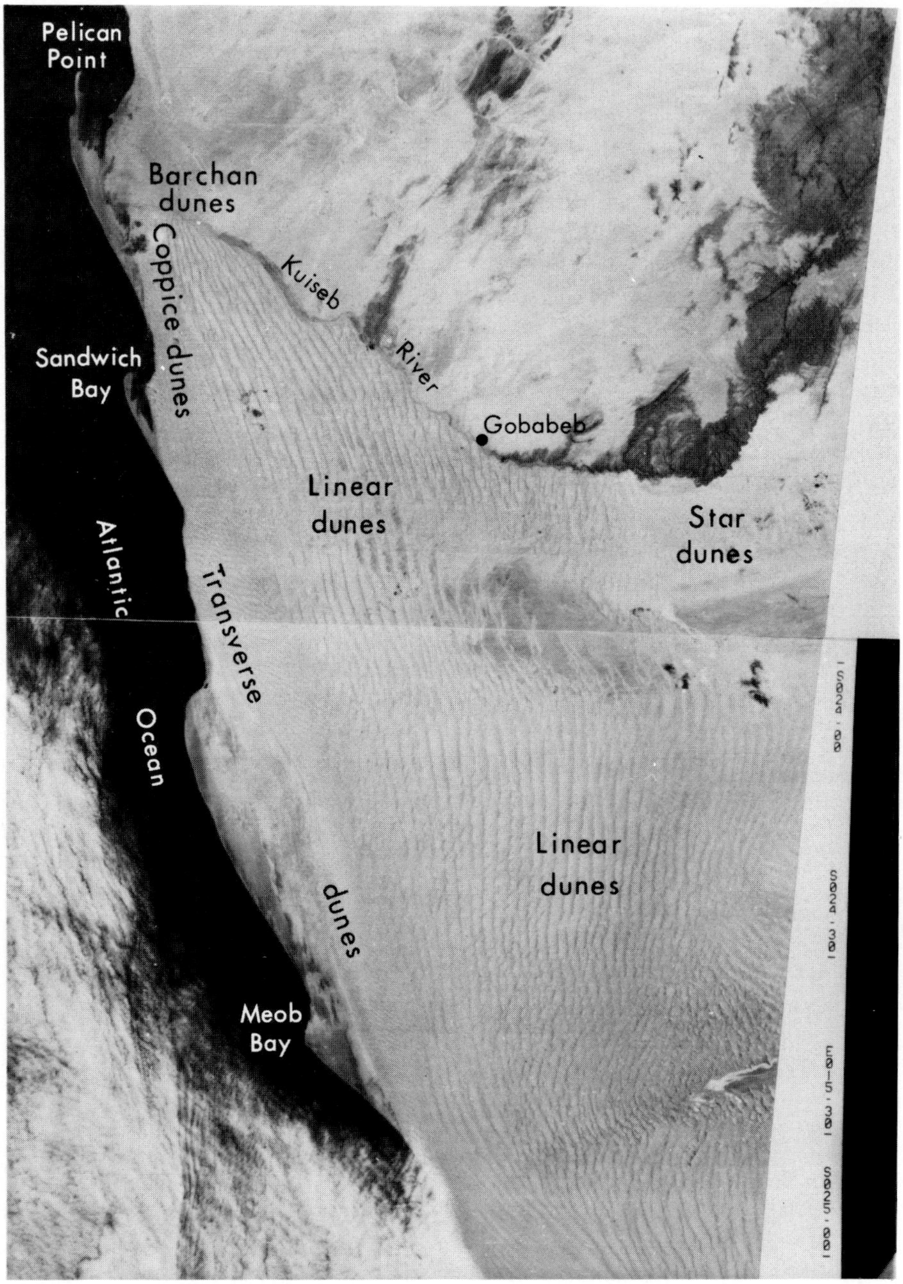

Figure 12. Landsat image of northern and western parts of Namib sand sea. The Kuiseb River crosses the picture, right to left, in upper part. North-south–trending ridges covering most of area are linear dunes. Atlantic Ocean is to left, bordered by east-west–trending, transverse dunes along coast.

they are also the most controversial and difficult to explain. These dunes are distributed across most of the interior desert area, 25 to 50 km inland from the Atlantic, to near the base of the Great Escarpment about 40 km to the east. They consist of long sand ridges oriented north-south, in general, and they extend from near Luderitz in the south about 155 km to the Kuiseb River in the north (Fig. 1).

Characteristic features of linear dunes are their sharp-edged crests, symmetrical cross-profiles with slipfaces formed alternately on one side and then the other, their elongation for kilometres in one direction, and a nearly equal distance between dunes that is occupied by relatively flat valleys or interdunes (Fig. 12). In general, the crest lines are sinuous and they rise and fall to form peaks and saddles (Fig. 13). Linear dunes, although common to most major deserts, have a number of different forms and are described under various names such as longitudinal dunes in Australia and seif dunes in north Africa. The Australian forms commonly have Y-junctions; Libyan Desert forms characteristically are relatively low and display zigzag patterns.

The genesis of linear dunes has been and remains one of the most difficult problems of interpretation in the realm of dune study. This situation is largely because of the difficulties involved in making direct observations both of the wind patterns involved and of the three-dimensional sand structures within the dunes. Many studies of dunes have attempted to explain the origin of these forms with the result that a large number of different hypotheses have been advanced and considered. In general, these hypotheses can be grouped into three main classes, but many other explanations that involve combinations of the basic elements or that explain only certain types of linear dunes in specialized environments are recorded.

Deflation Hypothesis

Probably the oldest hypothesis to be advanced in explanation of large fields of nearly parallel, evenly spaced, linear dunes is that of deflation. The long sand ridges were considered by Aufrere (1928) to be remnant features of relief resulting from erosion of the valleys between them. Similarly, in the Thar Desert of India, Blandford (*in* King, 1960, p. 102) considered the ridges to be residual sand rather than sand piled up by the wind, and the intervening valleys to be products of denudation. In South Australia, King (1960, p. 99) stated that deflation had "played a direct role in dune formation and in the excavation" of intervening playa lake basins.

For the Namib Desert, a variant of the deflation origin has been suggested by Barnard (1973, p. 10). He stated that the "longitudinal dunes of the central zone seem to have been initiated by processes of deflation."

As information concerning linear dunes has accumulated, the plausibility of their origin through deflation has decreased until it is now given little credence by most geologists. Critical evidence against it has come with the technique of trenching dunes to determine their internal structure. Typical dune cross-strata exposed in linear dunes (McKee and Tibbetts, 1964; Tsoar, 1978) have shown conclusively that these dunes were products of deposition and not of erosion.

Helical Roll Hypothesis

A second and relatively popular explanation for the origin of linear dunes is through the medium of longitudinal helical roll vortices. This concept seems to have originated with Bagnold (1953, p. 91) who stated that "cells take the form of pairs of long roller vortices with parallel horizontal axes so that particles of air in their passage downwind tend to describe helices." He further stated that helicoidal flow is caused by differential heating of sand surfaces and causes the "sand to become organized into vast parallel ridges"—the spacing between them ranging from 2 to 5 km.

In support of his hypothesis, Bagnold (1953, p. 93) made a comparison between wind records for a 10-year period and major dune directions in the Egyptian desert. Resultant wind directions showed little agreement with mapped dunes when calculated for the entire year, "but good agreement when hot months only were included." Support was given the helical roll hypothesis when Hanna (1969), a meteorologist, presented measurements from the atmosphere and described a mathematical model. He stated, however, that "proof will come when experiments with weather balloons are conducted" and when flows that are allegedly the cause of a stable dune alignment are demonstrated.

In the Namib Desert, the helical roll hypothesis was adopted by Besler (1977, p. 51, Fig. 4) to explain the typical linear dune development, but she presented no evidence in support other than favorable size and distribution of the dunes. The apparent lack of sufficiently strong winds today to develop these large ridges was countered by her with the conjecture that possibly much stronger winds occurred during the Pleistocene.

In opposition to the helical roll hypothesis is the conclusion of Tsoar (1978, p. 141) based on detailed studies of linear dunes for a period of several years: "In the present research, no proof was found that the longitudinal dune is formed by the helicoidal flow or by a unidirectional wind."

Dune Modification and Evolution Hypothesis

Some varieties of linear dunes seem clearly to have developed from, or to be the extensions of, other dune types. This process applies especially to simple dune forms that begin as deposits of unidirectional winds and later are modified by crosswinds. According to Bagnold (1941, p. 222), barchans and transverse dunes can be regarded as the

Figure 13 (this and facing page). Linear dunes with star dunes and reversing structures superimposed. Interdunes (streets) on both flanks, south of Gobabeb. Aerial photographs by E. Tad Nichols.

origin of the longitudinal dune. The concept that one of the horns of a barchan, on the side exposed to storms, may be elongated by nonconstant but strong winds was elucidated by him and has been documented and elaborated on by numerous other geologists (Cooke and Warren, 1972, p. 300; Tsoar, 1978, p. 131; Lancaster, 1980, p. 160).

A second variety of linear dune that probably developed from a simple dune form, and was later modified to another type of dune by changes of environment, is represented by the trailing arms of parabolic dunes. In Pakistan this type of dune is common where vegetation, resulting from the monsoom, anchors the dune arms to form linear ridges. Dune noses connecting the ridges commonly are destroyed and arms are stabilized. Such sand forms have been described by Verstappen (1968) and by Cooke and Warren (1973, p. 318).

Still another primary sand form to which some linear dunes have been attributed is the zibar. It is defined as a low dune, only 4 to 5 m above the interdune surface, and without slipfaces (Cooke and Warren, 1973, p. 309). It occurs mostly in interdune corridors and forms hard surfaces of bimodal sand. Zibar strips commonly occur at the beginnings of longitudinal dunes and are considered "frequently a source or origin" of these dunes (Tsoar, 1978, p. 23). "Lunettes or foredunes on the borders of playas" in the Simpson Desert of Australia are likewise considered starting places of linear dunes (Wopfner and Twidale, 1967).

Lee-side Accumulation Hypothesis

Some linear (longitudinal) dunes have been interpreted as the result of "wind shadow" development of sand accumulation to the lee of outcrops or other obstacles (Melton, 1940, p. 133). Examples given of such elongate forms are on the Moenkopi Plateau, Arizona, and in the Rann of Cutch, India, where the effective wind direction is narrowly controlled by physical barriers.

Seif dunes along the borders of the sand sea in Libya were observed (McKee and Tibbitts, 1964, p. 5) to "originate where vegetation and/or small rock masses form sand traps rising above the general surface." Such physical features, of course, only account for the beginning of a sand ridge, the maintenance and elongation of which clearly are functions of regional and local wind regimes.

Bidirectional Wind Hypothesis

A compelling explanation of linear dune genesis comes from several studies of this dune type in areas where the effects of bidirectional wind deposition have been critically examined. Probably the first geologist to promote this hypothesis was Bagnold (1953, p. 301) who suggested that in the Arabian desert repeated wind cycles created long dunes that were aligned, not with a persistent wind, but with the vector sum or resultant of all sand-moving winds in an area. He referred especially to storm winds, blowing diagonally, in addition to the prevailing winds.

The bidirectional wind hypothesis was tested on the Oregon coast by Cooper (1958, p. 56, Fig. 9) where "oblique" dune ridges trending N. 40° W., were shown to result from a combination of summer winds, N. 10° W., and winter winds, S. 35° W. He concluded that "crossing and recrossing the crest, the sand pursued a zig-zag landward course, following the trend of the ridge, thus contributing to its nourishment and extension. As ridges grew in bulk through absorption of the mobile sand, the interridge corridors suffered depletion; their floors were lowered, in places to deflation base."

Conditions essential to the development of an "oblique ridge pattern" (Cooper, 1958, p. 57) are (1) exposure to full force of both seasonal winds, (2) adequate supply of new sand, and (3) level space, not abruptly sloping, to extend itself inland (forward). These conditions probably apply to most if not all linear dunes that have been recorded. Furthermore, the proportions of the ridges, height versus length, seem to be functions of the angle at which the two principal winds intersect one another, and this angle may vary greatly for different regions. As stated by Cooper (1958, p. 63), if the angle is wide, massiveness is emphasized rather than extension; but if the angle is narrow, the effectiveness of the forward component is favored and the ridge will extend itself rapidly at the expense of its bulk.

In the Simpson Desert of Australia, additional support for the concept that linear dune ridges are similar in orientation to the trends of resultant wind directions has been furnished by Madigan (1946) and by Wopfner and Twidale (1967).

Direct evidence of deposition through bidirectional winds at an oblique angle is furnished by dune structure as manifested by the presence of opposing slipfaces. Characteristic features of most linear dunes are sharp crests with high-angle slipfaces on both sides, the result of an alternation in direction of windward- and lee-side deposits. The long-term record of such development is found in sections of cross-strata exposed by trenching normal to the dune crest as illustrated in Libya (McKee and Tibbitts, 1964) and in the Negev Desert (Tsoar, 1978, p. 104–107). Such structures are in contrast to the high-angle foresets of dunes formed by unidirectional winds, where cross-strata consistently dip in one general direction.

The recent detailed studies of Tsoar (1978) in the Negev Desert have done much to establish the dynamic and the structural evidence of linear dune development by winds from two directions. This work consisted of rigorous tests involving direct measurement of wind direction and velocity, tracing sand-grain movement by using fluorescent dyes, measuring with markers both dune advances and profile changes, tracing wind currents with smoke plumes, making

Figure 14. Linear dunes and interdunes, southwest of Gobabeb. Aerial photograph by E. Tad Nichols.

grain-size analyses, and conducting trench studies of dune structures.

Conclusions reached by Tsoar (1978, p. 131) include evidence that linear dunes are secondary forms developed from barchan, transverse, or zibar types and that additional sand furnished to linear dunes causes elongation, whereas with transverse dunes it increases height. Further, the studies by Tsoar (1978, p. 130) show two basic necessities for developing linear dunes: (1) a bidirectional wind regime with an angle not greater than 150° to 180°, and (2) a sharp profile with each side 16° to 20°, so a "flow of separation" will develop on the lee flank. He summarized his dune studies (Tsoar, 1978, p. 141) by stating that "the oblique cross winds are the main factor required for longitudinal dune dynamics and morphology."

Namib Desert Linear Dunes

Linear dunes of the Namib Desert sand sea are seen on Landsat images (Fig. 12) as very long, nearly parallel ridges that trend in a general northerly direction. All along the ridge crests are scallops or indentations which, at the great distance of Landsat, appear as short, thin cross-ridges or narrow diagonal bars projecting into the dunes from adjacent interdunes. At much closer range, as in aerial photographs (Fig. 13), the main ridges are seen to consist of a series of S-curves with embayments facing alternately east and west. In places, narrow curving arms from the dunes extend out into the interdunes; elsewhere, two dune ridges come together to form Y-junctions that terminate interdune passages.

Around the margins of large vleis or arroyos, most of the dune ridges terminate abruptly (Fig. 14); at the downstream or western ends of vleis, dune ridges seem to bend or arch around the terminus as though blown out in that direction. Marginal to the vleis (Fig. 14), also along the Kuiseb River on the northern edge of the dune field (Fig. 14) and at some other places where winds from several directions meet, complex dunes composed of stars on linear ridges are forming. Where bidirectional winds occur with nearly opposite directions of movement, reversing dunes commonly form along the linear crests.

The exceptionally large size of linear dunes in the Namib Desert is one of their outstanding features. In the area near Gobabeb, they average "80–100 m in height" (Barnard, 1973, p. 24). "A maximum measured height of 170 m" and a common range between 100 and 150 m is recorded by N. Lancaster (1981, written commun.). These figures are considerably more than the "25 m or less" for linear dunes in the Kalahari Desert, the 15 m for seif dunes near Sebhah Oasis in Libya (McKee and Tibbitts, 1964, p. 10), or the 10 to 35 m of longitudinal dunes in the Simpson Desert of

Australia (Wopfner and Twidale, 1967, p. 134). In general, "a typical longitudinal dune is 20–50 m high" according to Hanna (1969, p. 875).

Very large linear dunes of the Namib Desert type have been referred to as megadunes or as draas in some regions (Tsoar, 1978, p. 19; Cooke and Warren, 1973, p. 282, 313). No basic differences in structure or in geomorphic form are known to exist between the large and the small forms, however, so probably the same processes are responsible for the genesis of each. Differences are in degree rather than in kind. If this conclusion is valid, then the question is raised as to the reason for the extremely large size of the Namib Desert dunes.

Two principal explanations have been suggested for the basic dune pattern and the height of dune ridges in the Namib. One is that they are products of an earlier time (Pleistocene). Winds are postulated to have been stronger then than now, and possibly large helical roller vortices were active in the area (Besler, 1977, p. 51). Today, only modifications of the dune summits occur, so no appreciable migration or growth is taking place.

The alternative explanation is that both the linear form and the dynamics of ridge growth have remained essentially the same for a long time. The dune pattern, once established in the southern area through the influence of outcrop or other local controls, has gradually been expanded northward, and the dunes have grown larger under a bidirectional wind control. Whether the dunes became higher or advanced northward at any particular time and place depended largely on the wind regime and available sand supply at that stage of development. The angle at which opposing winds approached each other, according to this hypothesis, has basically controlled the form, size, and movement of the dune ridges.

As suggested earlier, the beginning of the dune pattern in the south may have resulted from plant growth or rock outcrop distribution. On the other hand, the beginning and migration may be the result of alteration from barchanoid-type dunes by the process originally described by Bagnold (1941, p. 222) and recently elaborated on by Lancaster (1980, p. 161). In support of this concept, sketch maps drawn by Lancaster (1980, Figs. 2, 3) from aerial photographs show linear dunes beginning to re-form from small barchans in an area (Tsondab Flats) where earlier linear dune chains had been disrupted and dissipated as a result of alluvial deposition.

Wind Regimes Associated with Namib Linear Dunes

Wind roses and sand roses compiled recently by Fryberger (1979, Figs. 229, 230, 231) for the dune fields of the Namib Desert, give some insight into the principal wind regimes and their relation to the linear dunes. At Aus, near the southern edge of the desert, three principal sand-

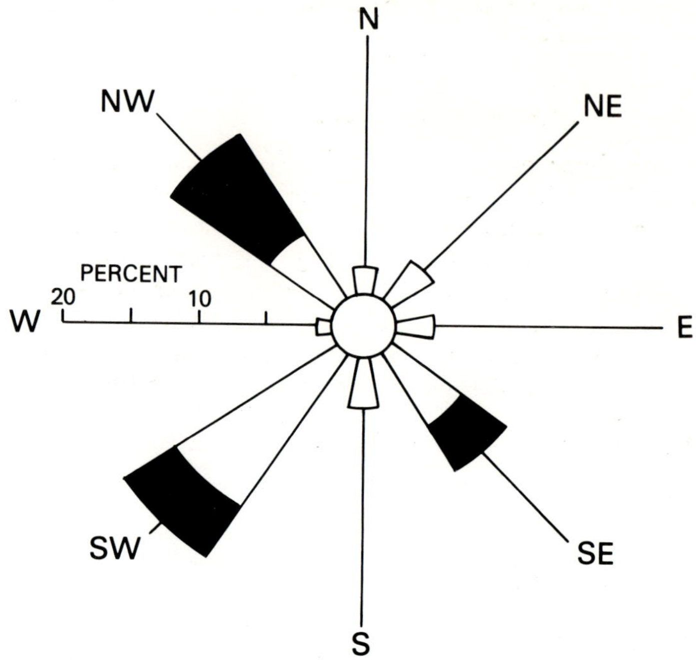

Figure 15. Wind rose for Aus, South West Africa, near fields of star dunes in the interior of the Namib Desert. Data were insufficient for computation of sand roses but indicate sand-moving winds from several directions; 1 cm equals 10% occurrence. Light areas, winds 14 to 27 knots (25 to 50 km/hr); dark areas, winds 28 to 40 knots (50 to 75 km/hr). Source of wind data is Royal Navy and South African Air Force (1944). From Fryberger (1979, Fig. 231).

moving winds are recorded (Fig. 15): northwesterlies with a maximum amount of strong wind (25 to 50 km/hr), southwesterlies with the greatest amount of total wind (25 to 75 km/hr), and southeasterlies with a moderate amount of all wind strengths. The predictable result of the wind pattern at Aus is a considerable development of star dunes in the area.

Along the Kuiseb River at the northern margin of the sand sea, wind data for two stations—Gobabeb in the center and Rooibank to the west—were compiled by Fryberger (1979, Figs 229, 230) in the form of sand roses. These sand roses are circular histograms that depict the amount of sand potentially moved by winds from various compass directions at given geographic localities. At Gobabeb, southwesterly winds are shown as very strong in October, and to be present but weak in November and March (no December data are available). A second effective wind direction—the berg wind—is from the northeast and east. It is strongest in June, July, and August.

The station at Rooibank in the northwestern part of the Namib dune field (Fryberger, 1979, Fig. 230), like Gobabeb to the east, shows two principal wind directions—southwesterly and northeasterly. Because in both localities the two winds are from nearly opposite directions, reversing dunes are numerous in the Kuiseb River area. Additional winds in some places have further modified these dunes into star types.

Structures of Namib Linear Dunes

Dip degrees and directions in cross-strata of linear dunes were determined by trenching. Generally, the trenches were dug both normal to and parallel to the dune ridge. They were located at the dune crest (Fig. 16), which is the most temporary and vulnerable part of the dune, near the base of the main slope on each side of the ridge, and on the basal shoulders or plinths, beyond the slipface boundaries (Figs. 17, 18, 19) that develop on many linear dunes (Bagnold, 1941, p. 229).

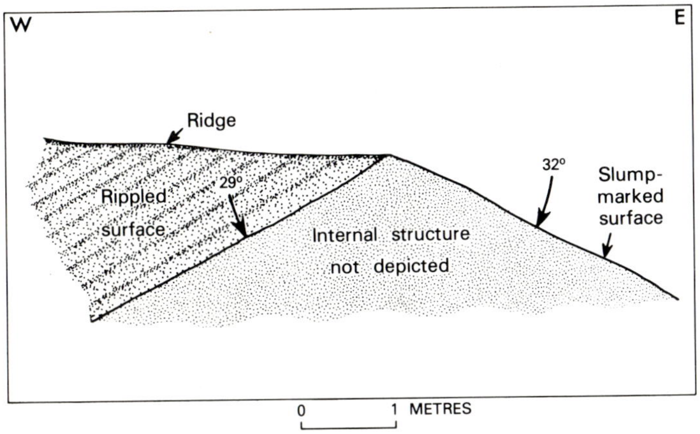

Figure 16. Dips of surface along crest of linear dune, 77 m high, between two star dunes, 7 km south of Gobabeb.

Cross-strata trends were determined for a number of linear dunes bordering the interdunes in areas south of the Kuiseb River, from the vicinity of Tsondab Vlei, westward to Narabeb Valley near the Atlantic Coast. The general structure as determined (Figs. 17, 18) consists mostly of high-angle foresets (about 28° to 35°) dipping toward the valley floor from both sides. Similar dip directions, but much lower angles (10° to 20°), are represented in the plinth on each side. Finally, very low angles (<8°) are preserved in the dune toe deposits along interdune margins.

Structure patterns seem to result primarily from deposition by avalanching and grain fallout on slipfaces that were developed alternately on each side of the dune ridges. This structure is interpreted to be the result of seasonal winds from two directions. Thus, the southwesterly winds in this desert (Goudie, 1972, p. 17; Logan, 1960, p. 5, 6) are in most places balanced by the strong berg winds from the plateau of central Africa to the east and northeast. The scalloped ridge lines so apparent in Figure 13 and the general linear pattern of dunes and interdunes seen in Figure 12 seem to be best explained as the product of bidirectional winds.

Textures of Linear Dunes

Sand samples taken from linear dunes near Rooibank and in the Nara Valley in the northern and western parts of the sand sea were screened and analyzed for grain size (Table 1). All but one of eight samples showed a mean value in the medium-size grade (¼ to ½ mm). The exception was in the fine-size grade (⅛ to ¼ mm). In general, sorting was "good" on the Payne scale in all samples, with only small amounts above and below the median. Other analyses of linear dunes in the Namib sand sea are presented and discussed by Lancaster (1981b, Table 2, Figs 3, 4). He described (p. 7) the movement of sands obliquely across the interdune corridors and the accumulation of the coarser grains along the plinths or dune bases, with the finer, better sorted sand continuing up the dunes..

Star Dunes

Where winds from more than two directions have contributed to dune growth, varieties of star dunes commonly

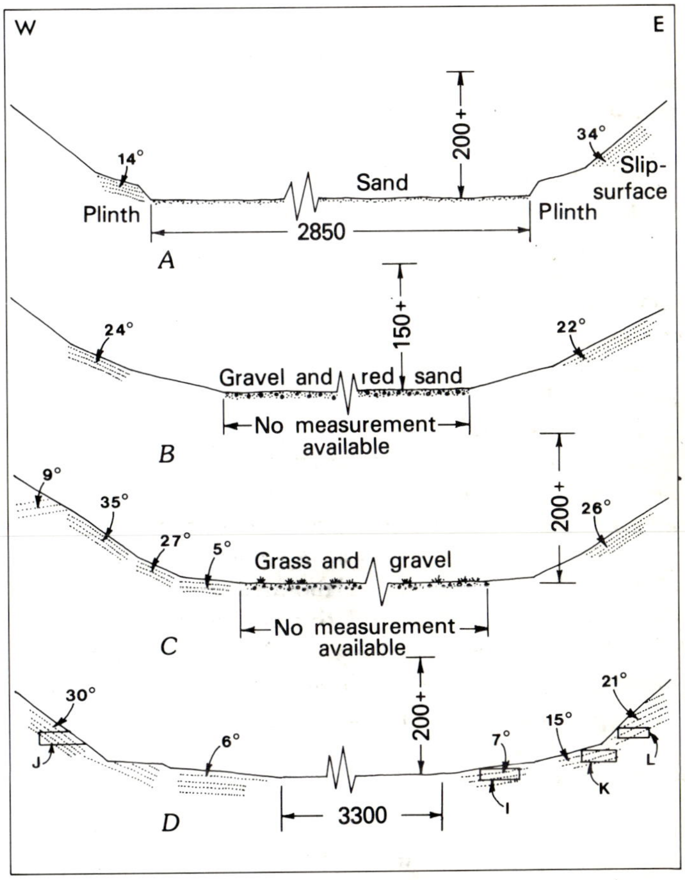

Figure 17. Cross-strata dips exposed in trenches on flanks of linear dunes on opposite sides of major interdunes. *A*, Nara Valley, 10.7 km south of Kuiseb River; *B*, Narabeb Valley weather station; *C*, 5.5 km south of Kuiseb River at Gobabeb; *D*, 6 km south of Rooibank. Horizontal and vertical measurements in feet.

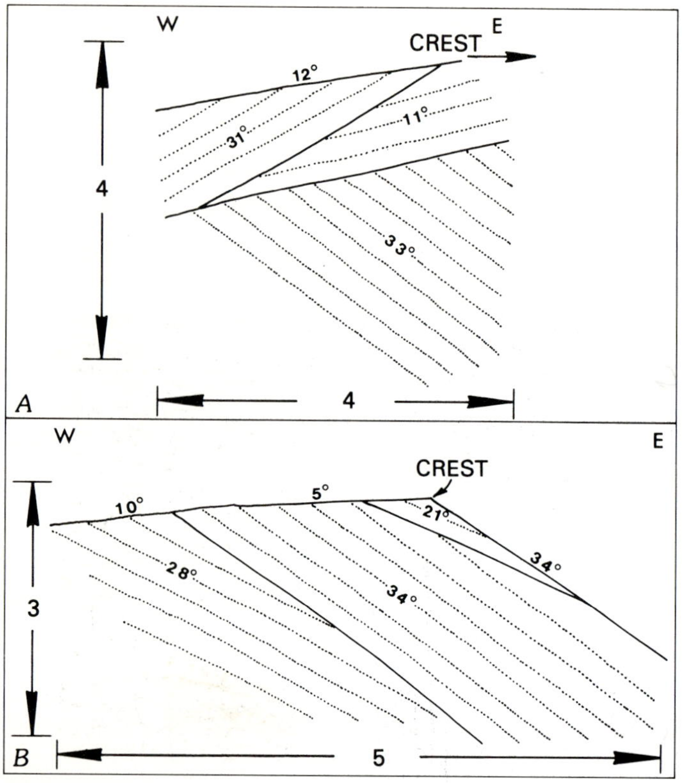

Figure 18. Cross-strata in trenches near crests of linear dunes, 2 km southeast of Gobabeb. *A*, west side of main dune; *B*, crest of secondary dune. Both illustrate dominant deposition toward east, but *A* also shows a reversal. Horizontal and vertical measurements in feet.

develop from or on top of the big ridges of linear dunes, with dominantly upward growth rather than forward migration. Such star dunes occur in numerous places along the Kuiseb River in the northern border of the sand sea. In this area, there are some northeasterly winds in addition to the strong winds off the central African Plateau (northeast, east) and those from the South Atlantic (southwest). Star dunes likewise have formed near the Great Escarpment on the eastern edge of the desert, especially along Tsondab Vlei and Sossus Vlei where various topographic features deflect air currents and cause complex wind patterns.

The star dunes commonly have three or four curving arms; a few have more. There seems to be no set pattern for form or arrangement of arms, and they extend in various directions. Some dunes are relatively small, about 15 m high for one south of Gobabeb, but others seem to have formed from large linear dunes or on top of linear ridges, and their points rise 90 to 150 m above the surrounding interdunes. One large star dune at Tsondab Vlei measured 106 m above its interdune base. Star dunes along the valley of the Tsauchab River are said to be about 300 m high (Barnard, 1973, p. 2, 5). Measurements by N. Lancaster (1981, written commun.) show a maximum of 279 m and a common height of ± 200 m.

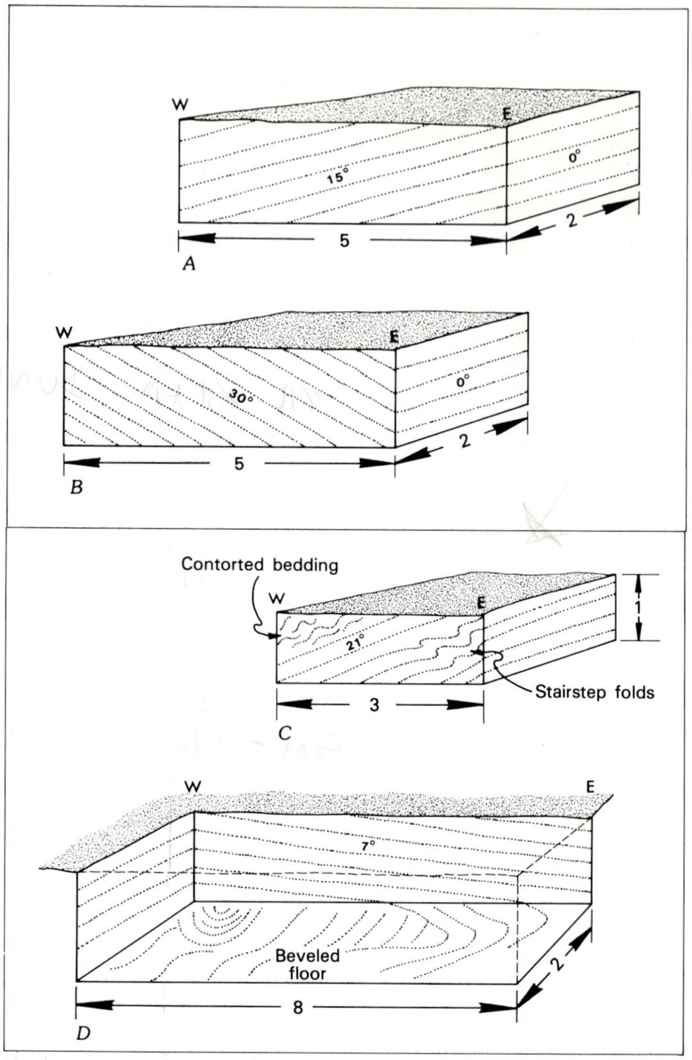

Figure 19. Block diagrams of structures in trenches of linear dunes and interdunes, 6 km south of Rooibank. Horizontal and vertical distances in feet. *A*, trench on west side of valley in plinth; *B*, trench on east side of valley, lower slipface; *C*, trench on east side of valley, contorted structure at base of slipface; *D*, trench in interdune, 0.15 km from base of plinth shows bevelled section of dune strata on floor.

Star dunes have formed in most of the large sand seas of the world—the Australian deserts being notable exceptions. Many names, mostly of Arabic derivation, have been applied to them locally. Rhourd (Aufrere, 1932, 1934; Smith, 1946; Stone, 1967; Wilson, 1971; Cooke and Warren, 1973) is perhaps the most common of these names. Ghoud is used in parts of North Africa (Capot-Rey, 1947; Capot-Rey and Capot-Rey, 1948; Mainguet, 1976); spelled ghord by Monod (1958). The names oghroud (Capot-Rey, 1947), pyramidal (Hedin, 1905; Holm, 1960) and conical (Petrov, 1967) probably refer to varieties of star dunes.

Grain size was determined for four star dunes considered typical of those along the northern margin of the sand sea

Figure 20. Incipient star dune, about 60 m high, developing from barchanoid type, 10 km north of Walvis Bay.

(Table 1). Results from these analyses were similar in all samples and showed mean-size grades to be in the medium class (¼ to ½ mm), and the next most common size was in the fine class (⅛ to ¼ mm). All other grain-size classes contained very small percentages, thus indicating good sorting. Sorting in star dunes was similar to that in linear dunes (Table 1), but in barchans and in large transverse dunes (Table 1), farther west near the coast, the percentage of fine grains was greater than that of medium size grains in most samples. Sorting was about the same (good) in all dune types examined.

Although most star dunes in the Namib Desert seem to be associated with or developed from the great ridges of linear dunes, there are exceptions. In the coastal barchan field north of Walvis Bay, for example, Nagtegaal (1973, p. 7) referred to "the barchans gradually losing their characteristic geometrical forms and developing, via transverse, into high, irregularly-shaped dunes." Such dunes, including those in the area south of Swakopmund, are approximately 60 m high and are found mainly at the northern end of a northward-migrating series of 16 or more, large, barchanoid dunes. They may be described as incipient star dunes where ridges tend to bifurcate, forming additional curving arms (Fig. 20), and some of them seem to be normal star dunes.

A star dune (Fig. 21) on the southwest margin of Tsondab Vlei is noteworthy because of its long sigmoidal arms, especially the arm that extends northerly. The dune is 106 m high, measured from the vlei surface on its east side. Profiles were measured at six places along its arm ridge. The steeper side ranged from 26° to 30°; the opposite side from 16° to 24°. Because the higher angled foresets dipped consistently in an east or southeast direction, that side was considered to be the lee side or slipface and the west or northwest to be the windward side at the time of examination. Evidence that it was not always this way is furnished by a trench made at the crest (Fig. 22). The section exposed in this trench (Fig. 22) shows a set of internal strata dipping east at a high angle (35°), overlain by another set that covers the bevelled ends of the older strata and dips in the opposite direction. Thus, a reversal of wind direction is recorded by the ridge structure with westerly winds dominant at the time.

Blowouts are associated with star dunes of the Tsondab Vlei areea and apparently are caused where curving arms of the stars block off wind movement and deflect currents from the normal regional trends to form pockets or inverted cones. Double stars (Fig. 23) or compound dunes of various types are also characteristic of this region, probably for the same reason, that is, the development of irregu-

A

B

Figure 21. Star dunes at Tsondab Vlei showing typical sigmoidal ridge arms.

Figure 22. Latex peel of structure section in trench at crest of star dune, Tsondab Vlei. East-dipping strata at 35°, overlain by surface layer dipping west to northwest; scale in inches.

lar curving arms that result from both local deflection of wind currents and from changes in regional wind directions with the seasons.

An attempt to classify star dunes (ghourds) by their distribution patterns or assemblages has been made by Mainguet (1976, p. 287) in north Africa. Patterns of groups as seen in satellite photographs enabled her to recognize five different assemblages (Mainguet, 1976, Fig. 3) based on presence of coalescence or of widely spaced occurrences, triangular or rectangular grouping, parallel or lozenge-shaped chains. Presumably, these patterns are well developed where regional winds are reasonably constant; in some parts of the Namib where stars are developing from other dunes, especially linear, the control of chain patterns is conspicuous. In other places, however, where conflicting winds from several directions meet, a lack of pattern and a chaotic distribution seem to result, as in an area southwest of Gobabeb (Fig. 24).

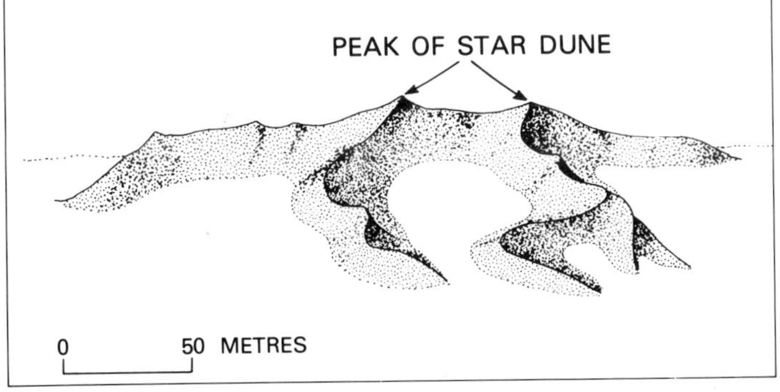

Figure 23. Double star dune, 6 km south of Gobabeb, bordering first interdune valley to east.

Because star dunes of the Namib Desert show great variety and have complicated patterns, three representative examples were selected for structural study. These dunes are all near the northern end of the sand sea, in the area of the Kuiseb River, and each is associated with one of the large linear dunes that extend as ridges in a north-south direction.

The three star dunes in which structures were examined are located as follows:

1. The first star dune is 17 km west of Gobabeb, on the south side of the Kuiseb River (Fig. 25). It illustrates derivation from reversing dunes.

2. The second star dune is 10 km southeast of Gobabeb, on the crest of a linear dune (Fig. 27). It shows characteris-

Figure 24. "Chaos" distribution of star dunes and others, between areas of linear dunes, southwest of Gobabeb. Aerial photograph by E. Tad Nichols.

tic branching at nodes or high points on the ridge and has compound star features.

3. The third star dune is on the southwest side of Tsondab Vlei and associated playas (Fig. 21). It is characterized by long sigmoidal, curving arms and apparently results from disruption of simple linear dunes. Blowouts are common features among the curving arms.

Star dune number one (Fig. 25) consists of two Y-junctions or points of bifurcation on a northeast-southwest trending ridge, so it actually consists of two three-pointed stars. All the arms are asymmetrical with steep slipfaces on the northwest, southeast, and northeast sides of three arms (from left to right). Strata on opposite sides of these arms dip 11° to 14°, 16°, and 18°, respectively, so probably represent windward surfaces. A trench dug at right angles to the crest line between stars (Fig. 26*A*) shows a reversal of dip direction. Figure 26*B* shows the northeast-trending arm and illustrates the steep side to the west (left) and gentle slope to the east (right).

Southeast of Gobabeb, about 10 km, star dune number two (Fig. 27*A*) rises about 18 m above a sand platform or base. It consists of a high point or peak with four arms or ridges extending out from it. The north- to northeast-trending ridge and the southwest-trending ridge are segmented, with changes in direction at the junctures. Large blowout depressions have formed in the lee of arms on the west side. Three trenches dug on the north, east, and southwest arms show foreset dips with high angles on the east and northeast sides and low angles in the opposite direction. A wide trench on the peak, exposing three-dimensions, shows high-angle dips to the north and west and low-angle dips to the east (Fig. 27*B*). These structure patterns are interpreted as meaning that deposition of the star dune was largely by winds from the west and southeast, but in the center from the south and east, indicating complex wind regimes.

Barchan and Barchanoid Ridge Dunes

Barchan dunes have steep slipfaces on their downwind sides and crescent shapes. Barchanoid ridge dunes are formed by the coalescing of numerous barchans in rows to form parallel wavy ridges (Fig. 28). Both types are formed by unidirectional winds—the barchan were there is a relative scarcity of sand available and the ridge type where a greater supply of sand is being furnished.

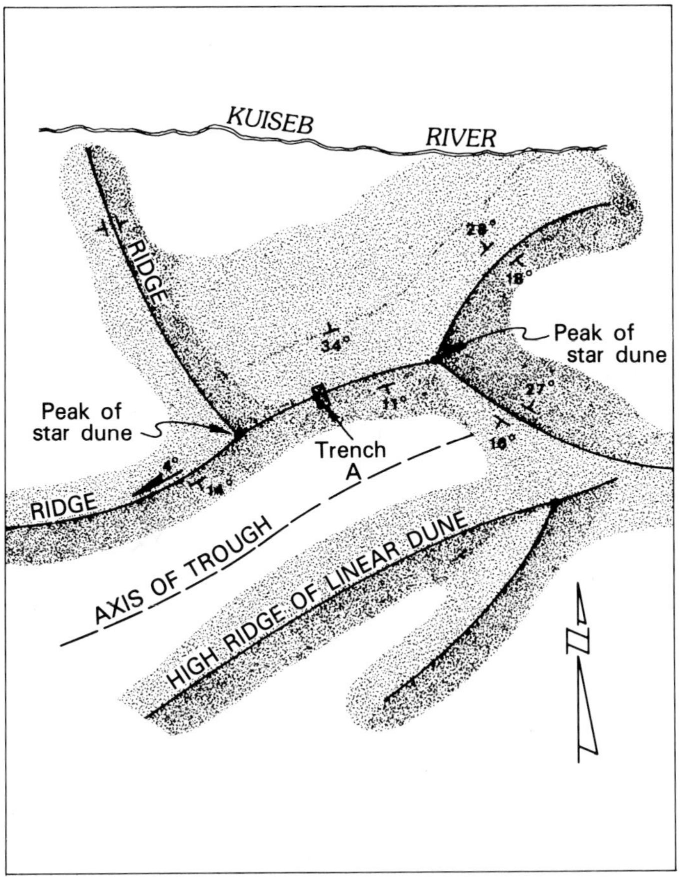

Figure 25. Plan view of reversing dunes and embryo star dunes, 17 km west of Gobabeb on south side of Kuiseb River. Distance between two peaks is about 50 m.

The part of the Namib best known for the developmen of barchanoid-type dunes is in the northwestern section, along the Atlantic Coast, from the Kuiseb River to the Swakop River. Dunes of this area have been studied in some detail by Nagtegaal (1973), and are discussed briefly by Logan (1960), Goudie (1972), Barnard (1973), and others. Small barchans occur farther south scattered along the Atlantic Coast within the belt of large transverse dunes that are likewise the product of unidirectional winds. Barchans are present, though not common, along the northern margin of the dune field on the south side of the Kuiseb River and at Tsondab Flats (Lancaster, 1980, p. 163).

Most of the barchan dunes of simple form, along the coast north of Walvis Bay, are relatively small, with heights of 4 or 5 m. Only a short distance inland, however, where much sand has accumulated, dunes consist of large barchanoid ridges with heights of 55 to 66 m. In this area, the barchanoids are progressively larger from south to north.

Orientation of the crescent-shaped sand bodies seems to be fairly consistent, facing in a north-northeast direction across much of the region, as recorded by Nagtegaal (1973, p. 7). This orientation results from southwesterly winds off the south Atlantic that affect the coastal area especially and cause dune migration toward the northeast (Fig. 29). Inland, south of the Kuiseb River in the vicinity of Homeb, barchanoid ridges that span an interdune between large parallel, linear dunes likewise face northeastward (Fig. 30), apparently the result of sand drift in that direction down the floor of the interdune valley. Barchanoid dunes in Sossus Vlei, formed by the northeasterly winter winds, are reported (N. Lancaster, 1981, written commun.), but they were not observed in the present study. The effects of such winds on other, higher dune types are recognized in many places, however.

In the northwestern barchan field, the interdunes as described by Nagtegaal (1973, p. 8) commonly are planar surfaces that separate dunes by several hundred metres. Farther north, the dunes are closer together and tend to coalesce. Many interdune deposits have crusts, mainly formed by halite, and surfaces are largely free of loose blown sand. In places, bevelled surfaces show the patterns of curving, truncated dunes of an earlier generation, and elsewhere thin layers of adhesion structures are prominent features (Fig. 5). All these features are being buried by dune foresets advancing across the interdunes.

Sand of the barchans and barchanoid ridges in the Walvis Bay area is predominantly of medium- (¼ to ½ mm) to fine-(⅛ to ¼ mm) size grades as shown in Table 1. Relatively few grains are represented in any other size grade, indicating the high degree of sorting. Heavy-mineral determinations, as described earlier under Sand Composition, are reported by Nagtegaal (1973, p. 12, Fig. 4) to show sillimanite to be by far the most common heavy mineral, with garnet, hornblende, staurolite, and epidote following in that order. Interdunes as represented by adhesion ripples had similar suites. Dunes and interdunes were intermediate between beach sands and Kuiseb River sands in mineral-distribution pattern.

Most dunes in the Walvis Bay area seem to be migrating inland. About 10 km north of Walvis Bay, a series of eight or more barchanoid dunes occurs with steep slipfaces toward the northeast. Each 70-m dune seems to be riding up the gentle windward side of the preceding dune, much as with the transverse dunes of Great Sand Dunes, Colorado, U.S.A. (McKee, 1979, Fig. 62). Near Swakopmund, at the northern end of this dune field, some of the dunes coalesce to form incipient star dunes (Fig. 20).

Internal structures of the barchanoid dunes of the northwestern part of the Namib dune field were examined by trenching at a few places (Fig. 30). They showed features similar to those of barchans recorded from other dune fields, with lee-side foresets of 30° or more and low-angle windward slopes bevelling the foresets of earlier stages. A barchanoid ridge 6 km southeast of the Kuiseb River near Homeb, connecting two large linear dunes, showed an intermediate set of low-angle strata dipping north, between

A

B

Figure 26. Embryo star dune and reversing dunes, 17 km west of Gobabeb. ***A*, structure in** trench in northeast-trending ridge connecting two star crests (Fig. 25); scale in inches; ***B***, water for trench being carried up arm of star dune, view northeast toward Kuiseb River.

A

B

Figure 27. Star dune, 10 km southeast of Gobabeb. *A*, trenching at dune crest; *B*, cross-strata exposed in trench at dune crest; scale in inches.

Figure 28. Row of coalescing barchan dunes at Walvis Bay with lagoon in foreground.

normal windward-side and lee-side strata (Fig. 30), indicating the nature of buildup in this barrier across the interdune valley.

Transverse Dunes

Transverse dunes are sand ridges that have relatively straight crests and that have slipfaces on the downwind side only. They are formed by unidirectional winds so are related to barchans; they differ from barchanoid ridges only in the regularity or straightness of the ridge line—the barchanoid type consisting of a row of crescents joined together. The transition from barchans, formed where relatively little sand is available, to the higher transverse dunes, with more sand, has been noted in the northwestern part of the dune field by Nagtegaal (1973, p. 7), and the observation that "transverse dunes are closely linked to barchans genetically" is recorded for that area by Barnard (1973, p. 10).

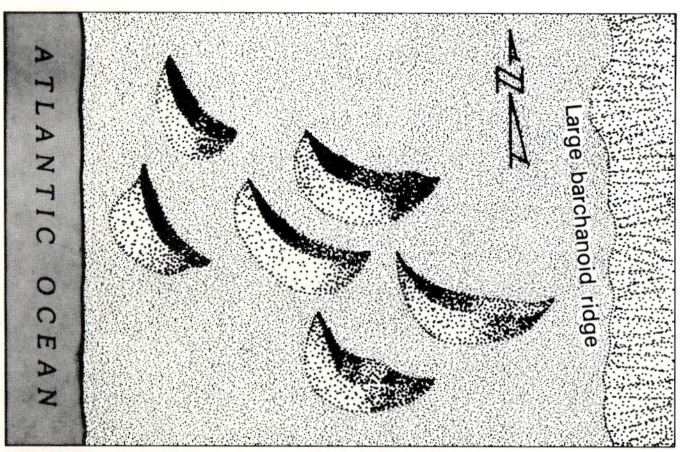

Figure 29. Small (5-m high) barchan dunes facing northeast; these dunes extend for several miles along west side of large (60 to 90-m) ridge of compound barchan dunes.

Two principal physiographic occurrences, representing distinct environmental situations, are represented by transverse dunes of the Namib Desert. One of these occurrences is bordering the Atlantic Coast where transverse dunes extend, in general, from Luderitz on the south to Swakopmund, north of Walvis Bay. A second type of situation characterized by transverse dunes is among the large linear dunes of the central sand sea. Here the transverse type commonly forms relatively low ridges across interdune valleys, and the ridges are oriented approximately at right angles to the linear dunes as first noted by Logan (1960, p. 136).

Coastal Areas

Coastal transverse dunes in much of the area south of the Kuiseb River attain considerable height (Figs. 31*A*, 31*C*),

Figure 30. Section across barchanoid ridge connecting two large linear dunes, 6 km southeast of Kuiseb River near Homeb; leeside to right; scale in inches.

and they extend inland for a number of kilometres. In general, they are oriented with steep slipfaces dipping toward the northeast, in contrast to the north-south–trending linear dunes that form most of the sand sea farther inland. The transverse dunes apparently reflect the prevailing direction of southwesterly winds off the South Atlantic, as described by Logan (1960), and are migrating northeastward. Sand from the coast seems to be the principal material represented in these dunes, as indicated by its general light color—much different from the red sand farther inland.

Examination of primary structures in several dunes in the vicinity of Sandwich Harbor confirms the direction of migration as postulated from the gentle windward slopes and steep lee slopes of the dunes. Cross-strata as seen in three dimensions in a trench at the crest of a transverse dune (Fig. 32) show steeply dipping beds (33°) on the northeast slope and low-angle strata at right angles along the crest. Color variations in these strata result from textural differences; irregular surfaces between some strata indicate minor erosional scours along the dune crest.

The long, relatively low-angle slopes to windward on transverse dunes of the Sandwich Harbor area, as seen from a lagoon bordering the Atlantic Coast, are shown in Figure 31*A*. Trench sections in the steep slipfaces on the northeastern side of each dune ridge show that dune migration is in that direction, but the considerable height of the dunes suggests an abundance of source sand and a slow rate of dune movement. Farther north near the coast, at about 15 km north of Walvis Bay, transverse ridge dunes seem to be developing from barchans and are much smaller than those at Sandwich Harbor. These dunes are similar in form and structure, forming a series of nearly parallel ridges (Fig. 33) facing northeastward. They seem to become progressively more compound-types downwind, to develop Y-junctions, and in places, to evolve into forms resembling star dunes.

Inland Areas

Transverse dunes of inland areas in the Namib sand sea are distributed through various interdune valleys between large linear dunes south of the Kuiseb River. They were examined at 6 km south of Homeb in the east, 2 km southeast of Gobabeb in the middle, and 22 km south in Nara Valley to the west. The ridge spanning the interdune south of Homeb, illustrated in Figure 34, shows the character of all these transverse dunes. It forms a low barrier with gentle dip to windward (south) and relatively steep dip to lee. Structures in a similar dune in the Nara Valley consist of buried foresets dipping north at 24°, with truncated tops, covered by a veneer of strata dipping south at 3° to 9° (Fig. 35). Thus, the ridge has migrated northward a considerable

A

B

Figure 31. Coastal transverse and barchanoid ridge dunes near Sandwich Harbor. *A*, extensive low-angle slope on windward side, facing lagoon on Atlantic Coast; *B*, crest of dune; trenching operations. *C, D*, steep lee sides of dunes seen from lagoon; views south along coast. Photographs by E. Tad Nichols.

C

D

Figure 32. Coastal transverse dune near Sandwich Harbor. Cross-strata exposed in L-shaped trench on dune crest; scale in inches.

number of metres. Other transverse dune cross-strata illustrating shifts in dune crest with northward advance of ridge were observed.

Interpretation of structures in the transverse dunes that have formed ridges across interdune valleys of the Namib is that they are caused by northward-moving winds that follow the valley floors between high linear dunes. They are advancing slowly, as testified by the considerable height of high-angle foresets forming the downwind margins. Furthermore, they are being fed with an abundance of sand, as shown by the tendency to develop veneers of low-angle laminae on the upwind sides where the wind has periodically been unable to move its load over the crest.

Analyses of grain sizes of inland transverse dunes show that in three samples examined (Table 1) all are dominantly fine grained (⅛ to ¼ mm). Similar analyses (Table 1) of sand from linear dunes that bound these transverse dunes show seven of eight samples to be dominantly medium-grained sand (¼ to ½ mm). The other sample was fine grained.

Reversing Dunes

Reversing dunes have been recognized for many years (Hedin, 1896, p. 267–268; Cornish, 1897, Fig. 7) and are reported from numerous widely scattered localities throughout the world (McKee, 1979, p. 103–107). These localities include both arid and humid areas at low altitudes and cold-climate areas at both high and low altitudes. In all these areas, the reversing dunes are formed from transverse or barchanoid dune types that normally are the product of unidirectional winds. Most reversing dunes, therefore, seem to result from infrequent but strong winds blowing from a direction essentially opposite that of the prevailing wind.

In the Namib Desert, reversing dunes are common in the northern part of the sand sea along the Kuiseb River valley (Fig. 1) and probably elsewhere, as at Sossus Vlei. In these areas, a reversal in slipface development occurs in the uppermost part of many dune ridges as a result of the strong winds blowing toward the west during the winter season. The resulting dune form resembles in most respects that of reversing dunes at Great Sand Dunes, Colorado, U.S.A., although those dunes are derived from transverse ridges, whereas the Namib-type dunes are derived from linear ridges.

A major difference between these dune types is that the transverse dunes, being dominated by a unidirectional wind, tend to climb up, one upon another, in shingle fashion, and

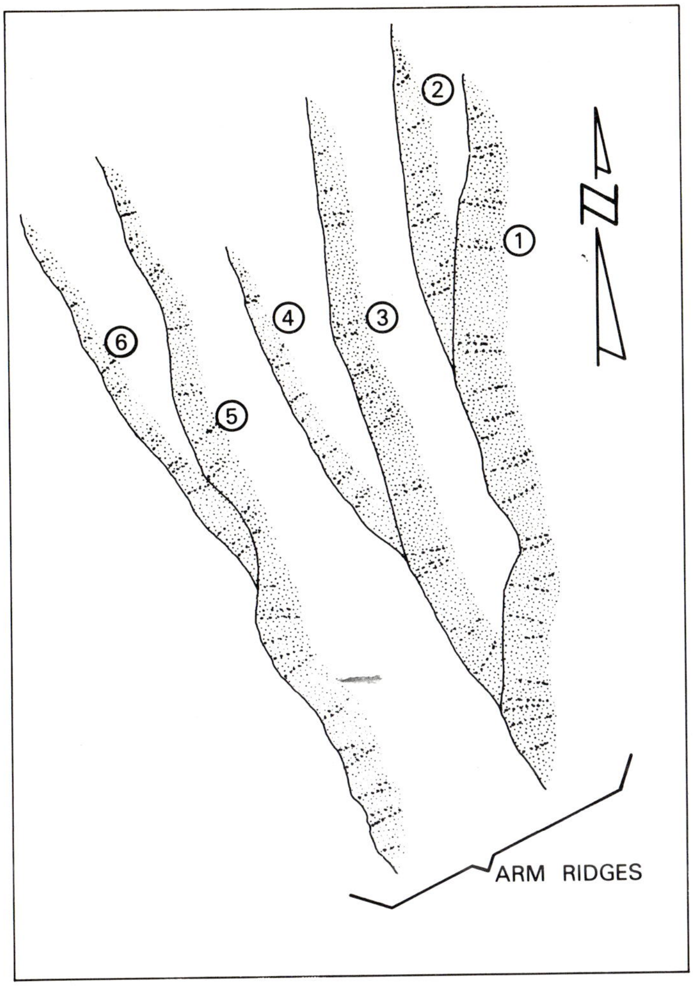

Figure 33. Plan view showing pattern of compound transverse dunes with branches and nearly parallel slipfaces; near coast 15 km north of Walvis Bay. Dune ridges 30 to 60 m high, slipfaces dip northeast. Approximate length of arm 1 is 150 m.

without any flat-lying interdune valleys between. In contrast, linear dunes of the Namib seem to be influenced almost equally by each of two wind directions, and so have a forward component of movement northward nearly parallel to the ridge axis; these ridges are separated by wide, flat interdunes. The most important factor responsible for differences in the two types of reversing dunes, both of which are the products of bidirectional winds, probably is the angle of convergence of the wind—in the one case from nearly opposite directions; in the other, at some large oblique angle.

A relationship seems to exist between normal reversing dunes and star dunes of the Namib. Both of these types are most numerous and best developed in areas near the sand sea margin an in proximity to vleis, large topographic features, and other features where the wind tends to be deflected in various directions. In brief, conflicting wind-direction vectors probably are responsible for most abnormalities in dune form. The crests of many such dunes or dune arms are extremely narrow with ripple marks normal to the crest line on the windward side and slump marks or avalanche patterns consipicuous on the lee side. Both sides may be relatively steep, approaching the angle of repose (Figs. 36*A*, 36*B*).

Dome Dunes

Dome-shaped dunes in the Namib Desert consist of low mounds of sand that occur uncommonly among various other types of dunes, especially along margins of the dune field, probably in areas of local changeable winds. One example, considered typical and located about 2 km south of Gobabeb, was selected for dissection and study. It was designated dune C.

Dune C, as shown in Figure 37, had an oval shape in plain view with a long dimension of 147 m and a width of 58 m. The height to the rounded summit was 3.6 m. This dune was on the eastern side of a large interdune corridor and was close to (55 m) the base of a large reversing ridge dune. Trenches dug in this dome at about the central high point (Fig. 38*B*) to expose structures in three dimensions showed that strata dipped with low angles in various directions from the middle (Fig. 38*A*), but no slipface, lee-side deposits with high angles were represented. The exclusively low-angle beds are interpreted as representing windward-side deposits, and their varied orientation suggests that the air currents must have deposited the sand from numerous directions.

The dome-shaped dune examined resembled in form dome dunes described at White Sands, New Mexico (McKee, 1966, p. 26), but differed from them in internal structure. The New Mexico dunes seem to have been formed from earlier barchan crescents in which strong one-directional winds bevelled and rounded their tops and modified their overall shapes. The genesis of Namib Desert domes is not readily apparent, but the location is postulated to have been determined by some surface obstruction, and the sand accumulation having no lee-side slipfaces probably resulted from changes in wind directions near the mouth of the wide interdune valley.

The texture of sand forming the dome-shaped dune under study was determined for two samples—one from the base and the other from near the crest (Table 1). In both samples the dominant grain size was medium on the Wentworth scale, which size is also characteristic of nearly all the large linear dunes (Table 1) and the star dunes (Table 1) sampled in eastern and central parts of the Namib dune field. In contrast, this grain size is coarser than in the transverse dunes of the Nara Valley to the west (Table 1) and in most of the barchanoid types sampled along the Atlantic Coast north of Walvis Bay (Table 1).

Figure 34. Low transverse dune (foreground) crossing interdune at right angle to high linear dunes on each side. Steep foresets of transverse dune are to right (north), 6 km south of Homeb.

Blowout Dunes

It is debatable whether or not the craterlike forms known as blowouts, which were developed in sand bodies by deflation through spiralling wind activity, should be classed as a type of dune. Their forms are developed nearly entirely by erosion, whereas most dunes are the products of simultaneous erosion and deposition. In blowouts, the sides dip steeply toward the centers to form cones and the sand from the interior is blown over the rims and beyond as indicated by the name.

Whatever the classification, blowouts are conspicuous features of many dune-sand deposits and form distinctive land forms. In the Namib sand sea, they occur in two principal environments. One of these is in coastal areas where blowouts are especially numerous among compound barchanoid ridges and embryonic star dunes (Fig. 39) with complicated forms. Areas of partial stabilization from either vegetation or salt crusts seem to be especially favorable for blowout dunes as in other coastal areas throughout the world.

In the interior of the Namib, many blowouts are formed along the sides of the great linear dunes that traverse the region from north to south, among the curving arms of star dunes and along the crests of reversing dunes. The complicated forms of these superimposed dune types seem to be caused by variable winds from different directions that have disrupted the linear ridge crests. Blowouts commonly develop during this disruption, as described by Logan (1960, p. 136).

Through the dry, interior dune region south of the Kuiseb River, blowouts are concentrated in many areas from Tsondab Vlei on the east to the vicinity of Gobabeb in the center, and at least to Soutrivier on the west (Fig. 1). No information is available on their distribution farther south, but they seem likely to occur wherever large variations in wind regime are disrupting dune forms.

In all localities where blowouts were examined they were formed between the sinuous, curving arms of star-type dunes; the denser and more complex the dune pattern, the more blowouts occurred. At 10 km southeast of Gobabeb, two small elliptical blowouts were next to each other on the west side of a 50-foot star dune that had four arms. At 5 km south of the Kuiseb River near Soutrivier, a series of three overlapping blowouts was noted on the southwest side of a low dune ridge that trended northwest-southeast (Fig. 40*B*) and was developing irregular arms. Partway down the sigmoidal arm of a large compound star dune (106 m) at Tsondab Vlei, two small arms branch from the main arm enclosing a pair of large blowouts (Fig. 40*A*). Thus, blowouts seem to

A

B

Figure 35. Transverse dune across Nara Valley interdune. *A*, preparing trench across dune crest; *B*, cross-strata in trench across crest of dune, foresets (R) dip 24°, windward strata (L) dip 3° to 9°, 2.5 cm veneer of sand on top. Pencil is scale.

A

be a not uncommon dune type even in the dry interior of the sand sea.

Grain-size distribution in a sand sample from inside the crater of a blowout dune at Tsondab Vlei is shown in Table 1. The mean value for this sample is in the medium grain-size class (¼ to ½ mm), which is similar to but slightly less than in the star dunes of the same area (Table 1).

Coppice Dunes

The terms "coppice dune" and "shrub-coppice dune" are widely used in referring to small sand hummocks or mounds that have accumulated around plants, mostly in lightly vegetated areas. Such dunes commonly form in coastal areas (Cooper, 1967, p. 37) and also in semiarid interior regions where plants may partially anchor them (Melton, 1940, p. 135). Active coppice dunes move with the prevailing regional winds.

In the Namib sand sea, shrub coppice dunes are well developed in an area roughly 10 km south of Walvis Bay near the mouth of the Kuiseb River (Nagtegaal, 1973, Fig. 1). Farther south along the coast are high transverse dunes, and farther north is a field of scattered small barchans. The shrub coppice dunes form widely spaced mounds, 3 to 6 m high (Fig. 41). Many of these mounds have cross-strata on one or more sides conspicuously exposed as a result of etching by the wind with partial stabilization by plants.

Structure patterns within the coppice dunes of the Namib coastal area include large sets of cross-strata dipping at or near the angle of repose (Figs. 42, 43). They are interpreted

B

Figure 36 (this and facing page). Crest lines of linear dunes in areas of reversal. *A*, near top of embryo star, Tsondab Vlei; *B*, reversing ridge along crest of linear dune, southeast of Gobabeb.

as having formed by sand avalanching on lee sides. In some dunes the branches of plants are buried in sand, oriented crosswise to the laminae; trough-structures were noted in the upper part of some mounds (Fig. 43).

The high-angle foresets in the stratification of coppice dunes seem to indicate a predominating northward wind movement in this area, but the reversals probably show the results of alternate northwesterly winds from the north Atlantic or possibly winter winds from the northeast. Sand accumulating in these dunes apparently eventually feeds the barchan dunes farther north.

Interdunes

General Features

The interdunes or valleys between dunes are an integral part of any dune system. As is readily apparent on aerial photographs and Landsat images (Figs. 12, 13, 14), interdunes constitute a major percentage of the area covered by a dune field and occur in a large variety of forms, depending on the type of dune with which they are associated.

By far the most abundant and most characteristic interdunes in the Namib sand sea are the wide, parallel valleys between the major linear dune ridges. These valleys are locally referred to as streets and are remarkably straight for long distances, extending for at least 50 km in a general north-south direction. A total number of interdune valleys, separated by relatively narrow linear dunes, ranges from 27 to 35 between the Great Escarpment in the east and the Atlantic Coast on the west, a distance averaging about 100 km. Gradual changes in orientation, thickening and thinning of the valleys, and actual termination or subdivision of the valleys are conspicuous features of these interdunes in various parts of the dune field (Fig. 12).

Aside from the straight parallel corridors between linear dunes of the interior, certain distinctive features are locally associated with the ridges. Where two dune ridges come together at a low angle they form a Y-junction—a feature that occurs in many places a short distance south of the Kuiseb River. In various places south and west of Tsondab Vlei, the main sand ridges have developed numerous subparallel short ridges that are en echelon, and along margins of the vleis, star dunes occur in rows atop the linear ridges. Finally, a characteristic westward-extending bulge, not yet explained, has developed both in linear dunes and in interdunes on the downstream side of each vlei.

Major differences in the plan-view shape of interdunes are the result of variations in dune pattern. Barchan dunes and barchanoid ridges, for example, because of their crescentic forms, partially enclose their interdune surfaces and tend to develop pockets or basins as they migrate with unidirectional winds. Transverse dunes that also advance in a single direction, but with relatively straight ridges, likewise have narrow interdunes that are continually being buried on the downwind side of each dune and deflated to windward as the migration takes place.

Unfortunately for an understanding of dune development and history, relatively little is known concerning the geometric form of various interdune types. The depth or third dimension of Namib Desert dunes is not known from direct observations except along a few border areas, so as yet it is not possible to determine how many cycles of dune advance and dune burial are represented in the main sand body.

Surface features of Namib Desert interdunes are, of course, easy to examine and have been recorded from many localities. In general, the interdunes can be classed in two categories: dry interdunes of the interior and wet or intermittently wet interdunes of the coastal areas. In various Namib interdunes, the following features can be recognized: (1) lag gravel and other residual materials of deflation; (2) largely grass- and shrub-covered sand; (3) bevelled cross-stratified sand; (4) flat-bedded or structureless sand with or without surface ripple marks; (5) bed rock with remnant hills or bevelled flats; (6) sebkha or playa deposits with adhesion ripples, salt crusts, or black mud; and (7) lakes or vleis with carbonate precipitates.

Lag Gravels

Surfaces containing rounded gravels of various sizes and concentrations form interdune floors in many parts of the Namib sand sea. They were examined by the writer in four principal areas, from east to west, across the desert south of the Kuiseb River. Related gravels had been studied earlier in some detail by Logan (1960, p. 91) who recorded their diameters as commonly "less than ¼ inch" but, in places, considerably greater. Their areal coverage is stated to be "no more than ten percent of the surface," in general, but to cover as much as 50% locally where the pebbles are coarse.

The interdune gravels commonly form a desert pavement of highly polished and wind-etched pebbles, composed mostly of milky quartz, yellow quartz, and other durable

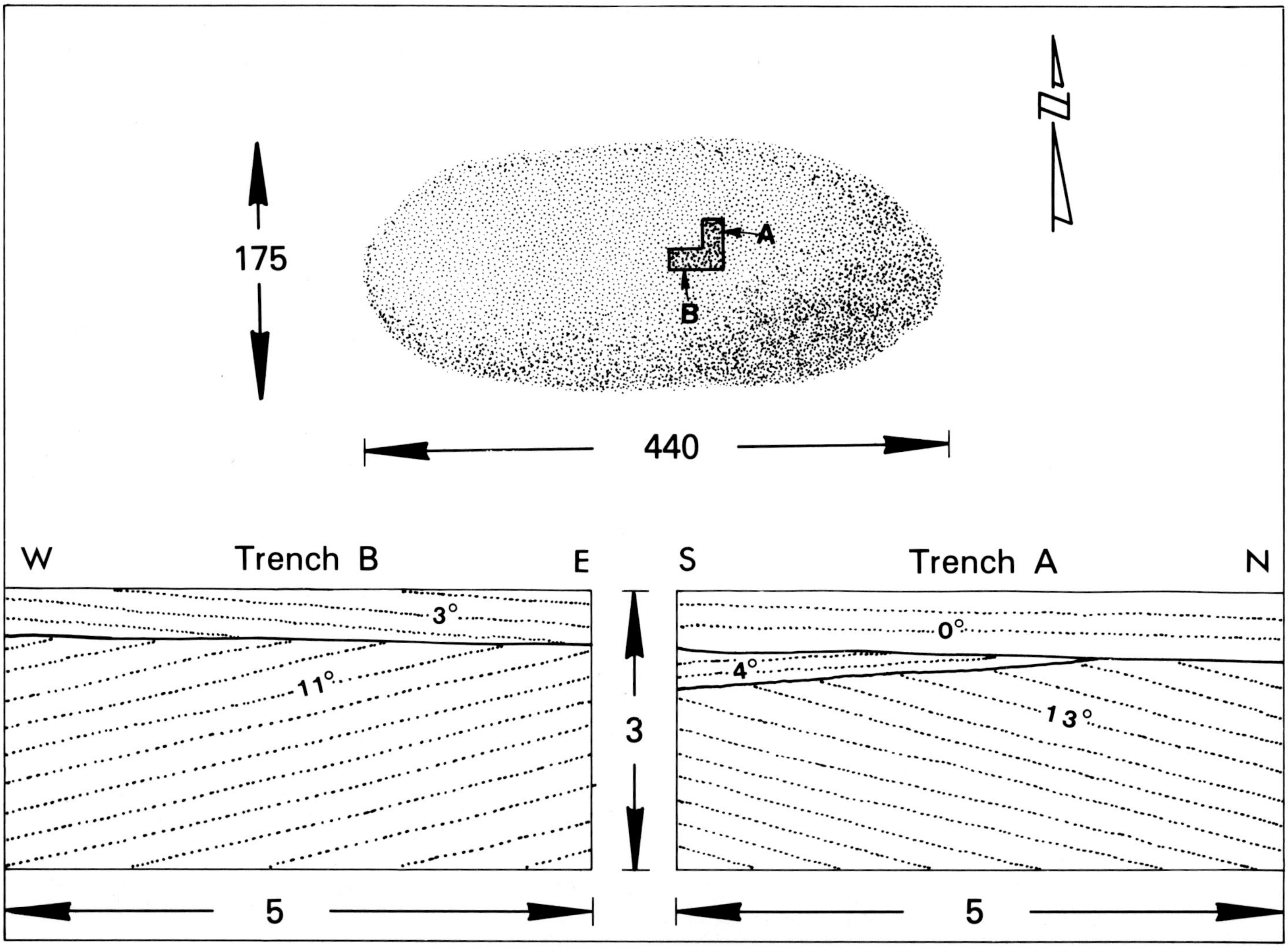

Figure 37. Dome dune, 3.6 m high, 2 km south of Gobabeb. *A*, plan view with location of trenches on crest; *B*, vertical sections in trenches A and B showing stratification.

A

B

Figure 38. Dome dune, 2 km south of Gobabeb near east margin of interdune. *A*, stratification near gently arched crest of dome; scale in inches; *B*, preparations for trenching at dune crest, view west across interdune.

Figure 39. Blowout in coastal dunes south of Swakopmund. Photograph by E. Tad Nichols.

rock types. They probably were originally deposited as alluvium and were distributed by streams off the Great Escarpment that lies to the east (Logan, 1960, p. 98). They are believed by Selby (1976, p. 5) to form a lag veneer, and he suggested that they were left at various levels by rivers that were incising the platform in several stages. An explanation that favors the concept of a "deflation of fine materials to leave a coarser residue at the surface" is suggested by Goudie (1972, p. 26).

Support for the deflation concept of concentrating gravels on the interdune surfaces seems to be furnished by trenches dug to depths of 3 ft and 4 ft, respectively, at 2 km south of Gobabeb and at Tsondab Vle. The scattered distribution of gravels with matrices of sand and silt in both of these sections strongly suggest the likelihood of concentrated gravel patches left after general sand removal.

Characteristic features of the gravels are the rounding of many into "tear-drop or more elongated dog-tooth form" (Logan, 1960, p. 92). Elsewhere, Logan (1960, p. 91) attributed the rounding to water which seems likely in most of the larger pebbles (Fig. 44*A*). In some areas, other types of clasts such as fragments of ostrich egg shells and calcitic molds of termite tubes are associated with the lag gravels and scattered among them (Fig. 44*B*).

Sandy Surface Interdunes

In the major linear interdunes of the Namib sand sea, which have widths commonly of 1,000 to 1,300 m, sand of unknown thickness forms a relatively flat or gently undulating floor in most places, especially in the interior parts (Figs. 2*A*, 2*B*). In some localities, chiefly along margins of the dune field, sand is mixed with lag gravel scattered randomly. Some interdunes have a sparse to fairly abundant growth of low shrubs or grass—its distribution controlled mainly by the recency and amount of precipitation.

Sand samples from 18 places in the dune field of the Namib Desert ranged in size from dominantly coarse grained (½ to 1 mm) to dominantly very fine grained (1/16 to ⅛ mm) Table 3). Ten of these samples, or approximately half, had mean diameters in the fine-size grade (⅛ to ¼ mm), and five samples had mean diameters in the medium-size grade (¼ to ½ mm). Thus, interdune sand shows more variability within the sample, as well as more variability throughout the region, than does the associated dune sand, as indicated by results of the relatively limited sampling done. Despite this variability, the sorting of the interdune sand is mostly fair to good on the Payne scale, exceptions being where lag material gives bimodal sorting.

A

B

Figure 40. Blowout dunes. *A*, deep overlapping blowouts formed among arms of large star dune at Tsondab Vlei; *B*, blowouts on southwest side of low dune ridge trending northwest-southeast, 5 km south of Kuiseb River near Soutrivier.

A

B

Figure 41 (this and facing page). Coppice dunes consisting of sand hills formed around shrubs, north of Sandwich Harbor on Atlantic Coast. *A*, numerous small mounds; *B, C*, large coppice dunes.

C

The dominantly sand interdunes commonly show only flat or ripple-marked surfaces with minor irregularities resulting from vegetation mounds or from deflation scours. With trenching, however, structures of bevelled cross-strata sets commonly are exposed, thus indicating that the structureless surface sand is only a veneer in such places and has been reworked locally.

The thickness of the dune sand beneath the surface and the number of generations of dunes that are represented below the interdune floors are as yet unknown quantities, except around dune margins where outcrops or other evidence of near-surface bed rock are apparent. In central parts of the sand sea, where regional downwarping may have occurred, possibly more than one dune sequence is buried, and these, if recognized, would throw much light on regional history. In contrast, evidence of a stable platform of ancient crystalline rock like that north of the Kuisab River would indicate that the history of the sand sea is essentially limited to accumulation of those deposits exposed today. The answer to this problem probably can be determined best by drilling in a few carefully selected places.

In the synclinal basin of the Libyan Desert of northern Africa, drilling has demonstrated (McKee and Tibbitts, 1964, p. 5) that in places at least 300 m of dune deposits are buried beneath the surface, yet in many other regions (Wilson, 1973, p. 103, Table 2), large modern dunes are forming on or near a bed-rock surface.

Minor features characteristic of large, dry-sand interdunes are represented in the interior parts of the Namib Desert by wind-ripple marks that typically have low crests and high indices (Fig. 2*C*), scattered shrub or grass mounds (Fig. 2*D*), plant root molds or carbon residues, grain-lag concentrates, and bioturbation traces, which are mostly termite tubes.

TABLE 3. SUMMARY OF MEAN GRAIN-SIZE DISTRIBUTION IN 18 RANDOM SAMPLES OF INTERDUNE SANDS

Grain size	Total of mean values
Coarse	2
Medium	5
Fine	9
Very fine	2

Bed-rock Hills in Interdunes

Interdune areas of the Namib sand sea include in some places, especially along the eastern and northern margins, exposures of bed rock that give insight concerning what types of rock and topographic forms might be expected to underlie much or most of the dune area. These rock outcrops suggest that a continuation of the plain or platform exposed north of the Kuiseb River may extend under the sand deposits for a great distance, as described by Logan (1960, p. 133)

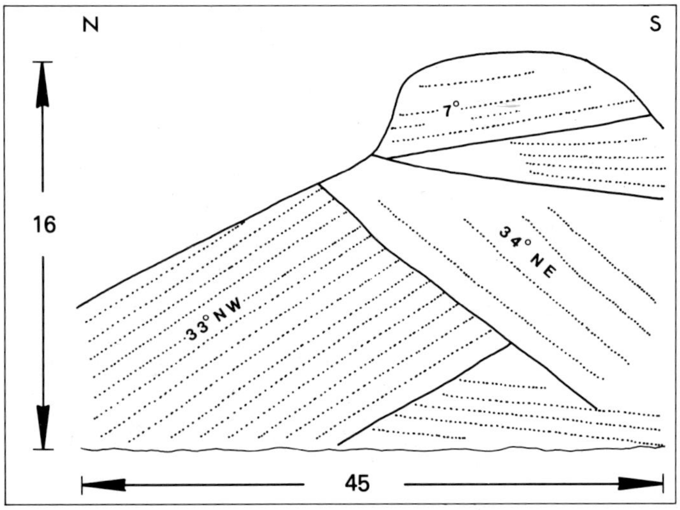

Figure 42. Structure section in coppice dune near Atlantic Coast, 28 km north of Sandwich Harbor. Cross-strata foresets near angle of repose in two directions.

and by Goudie (1972, p. 25). This conclusion has not yet been confirmed by drilling nor by sections exposed through dissection of the sand mass; however, "occasional high rocky mountains protruding from beneath" in the area of the great sand dunes as reported by Logan (1960, p. 133) are suggestive of such a situation. In many places, steep-sided, residual hills that rise abruptly from the surrounding plain or rock pediment, and that commonly are referred to as inselbergs, are prominent features. South of the river, small rock mounds and some relatively large ones project through the dunes at many localities and seem to mark a northern terminus of the large linear dunes or sand ridges. Examples, formed of granite, are conspicuous at Gobabeb (Figs. 45*A*, 45*B*), and ridges of schist project through the dune hills near Homeb (Fig. 45*C*). At Soutrivier, southwest of Gobabeb, they form hills 10 m high in the interdune.

In addition to remnant granite hills that rise above dune surfaces, residual bed rock of various types, including some relatively nonresistant varieties, is exposed as interdune floors at some localities. Sandstones, referred to as old red sands by Selby (1976, p. 6) are described by him as "planated but not removed by river erosion." Such bright red sandstones[1] form cliffs and residual ledges bordering Tsondab Vlei (Fig. 46*B*), and they are belived to be a principal source of the brilliantly colored dunes of that area. On the floor of the interdune at Soutrivier are extensive bevelled surfaces of cross-stratified red sandstone. The truncated tops of foresets have strikes that are largely northeast-southwest and indicate a number of separate sets with various orientations (Fig. 46*A*). Some have crescentic forms and others are straight; all were formed as large-scale cross-stratification, but whether they were water-deposited or eolian was not determined.

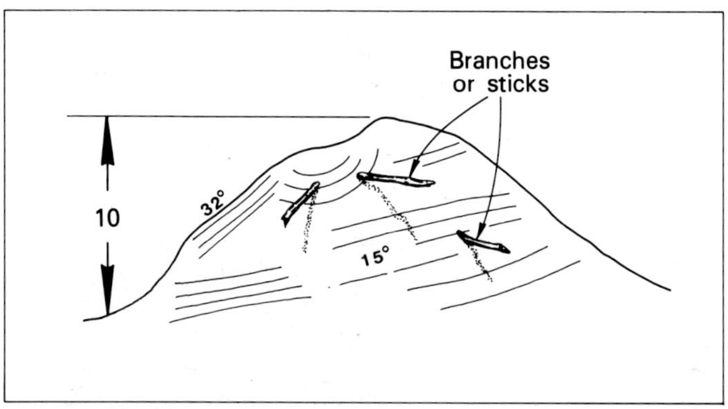

Figure 43. Sketch of coppice dune with trough structures in upper part and planar foresets below; partly buried branches of shrubs throughout, 15 km north of Sandwich Harbor. Vertical distance in feet.

Floors of Bevelled Cross-strata

Surfaces of interdunes that display the truncated edges of underlying cross-strata have been recognized in the dune fields of many regions throughout the world, especially where unidirectional winds prevail (McKee, 1966, p. 51, pl. VII). In places such as White Sands, New Mexico, and Padre Island, Texas, U.S.A., the traces of former barchan dunes occur as festoons of curving lamina on horizontal surfaces, and indicate course of migrating dunes by orientation of the arcs etched out on the surfaces.

In South West Africa, patterns of truncated cross-strata are common on interdune floors in the barchan dune field along the coast north of Walvis Bay (Nagtegaal, 1973, p. 8). The bevelled foresets of former small (about 5 m high) dunes show by the direction that the arcs face how at some earlier time barchans moved landward and northeastward across areas now interdune. Elsewhere, the traces left by migration of transverse dunes are recorded by relatively straight parallel ridges etched from weakly cohesive strata.

In wide interdunes of the Namib interior, between giant linear dunes, the complex patterns of bevelled cross-strata appear in some places as the traces of former dunes. About 6 km south of Rooibank, a trench dug in the interdune floor about one-third of the distance across the valley from its east side showed a complex pattern of both linear and arcuate laminae apparently representing the truncated arms of some earlier dune that once stood in that place (Fig. 47).

[1]A copy has just been received of the report on "a distinct lithologic unit" referred to as the Namib Sandstone (Besler and Marker, 1979, p. 155). This formation is described as a red sandstone that underlies much of the Namib Desert west of the Escarpment and seems almost certainly to be the same rock referred to above as a source of dune sand. The age of the Namib Sandstone and its derivation are said to be an open question (Besler and Marker, 1979, p. 159), though it is considered to be "relatively young" and "temporally distinct from the Pleistocene and recent dune sands that overlie it."

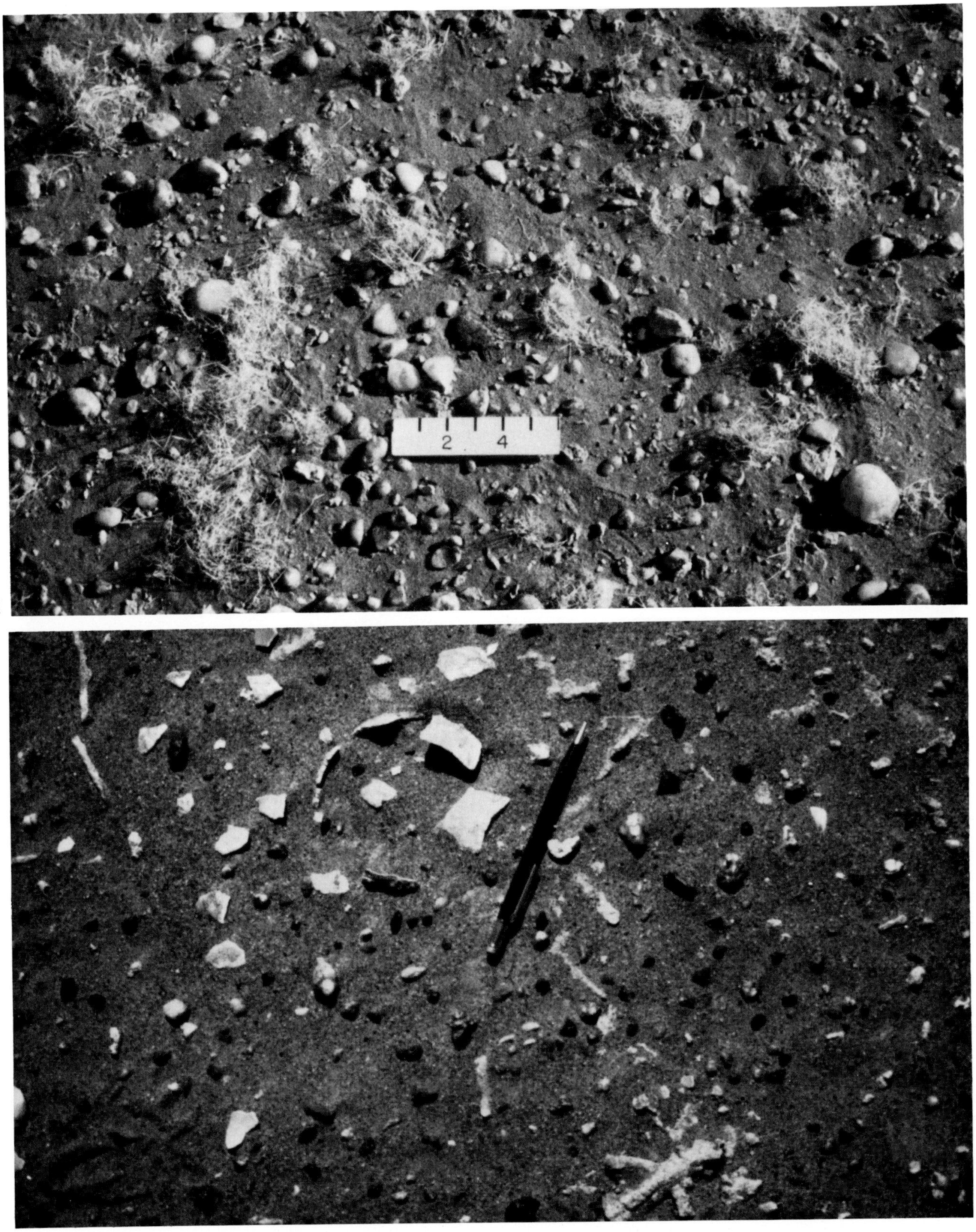

A

Figure 44. Gravel deposits on interdune surfaces. *A*, 3 km south of Kuiseb at Gobabeb, scale in inches; *B*, at Narabeb weather station. Deposit in *B* has fragments of ostrich shells and calcified molds of termite tubes. Black pencil is scale.

A

B

Figure 45 (this and facing page). Bed-rock hills in interdunes. *A, B*, granite outcrops (center) projecting through dune south of Kuiseb River near Gobabeb; *C*, outcrops of schist partly covered with drift sand, Homeb on Kuiseb River. Photographs by E. Tad Nichols.

C

Other examples of bevelled cross-strata from former dunes are represented in interdunes south of Gobabeb.

Areas of No Interdunes

The width of interdunes in various sand seas apparently differs greatly, but the factors controlling this width do not seem to be well understood. Where the wind is dominantly unidirectional, there probably is a relation between the amount of sand available and the ability of the wind to move it. Physical barriers in the form of mountains or cliffs, therefore, may either stop or retard the movement of migrating dunes.

Great Sand Dunes in southern Colorado, U.S.A., is an example of a dune field in which southwesterly winds have moved sand against a great barrier, the Sangre de Cristo Range to the east. As a result, large sand ridges have been piled up, one behind another against the steep flank of the mountain range. One dune ridge seemingly is climbing up the windward side of the preceding ridge, and so a shingle effect of overlapping dunes has developed. Unlike dune ridges of similar type on a flat, unobstructed terrane, the space between ridges has been greatly lessened so that interdunes with horizontal strata are absent, and the steep lee slope of one dune actually terminates on the lower angled windward slope of the preceding one.

In the Namib sand sea, a situation very similar to that of Great Sand Dunes occurs among the large transverse dunes along the Atlantic Coast south of the Kuiseb River. The high dune ridges in the area of Sandwich Harbour form ridge after ridge overlapping in a northeastward migration pattern (Fig. 31*C*). Unlike the dunes at Great Sand Dunes, however, no mountain barrier is present to account for the termination of the downwind movement, or the apparent piling up of one dune on the preceding one. Possible factors involved are (1) the large source of new sand continuously furnished along the coast, and (2) the inability of the wind, as it weakens inland, to move this load rapidly. Another probable factor is the effect of the strong counterwinds from the east that must be progressively more intense inland from the coast and that are supplying red sand from the east.

Sebkha Interdunes

In the barchan dune field south of Walvis Bay in the northwestern part of the Namib sand sea, many of the interdunes consist largely or entirely of sebkha deposits. Many of them exhibit on their surfaces the curving structures of bevelled cross-strata previously described; elsewhere an abundance of adhesion ripples forming round or oval humps (Fig. 5) occurs, and in many places considerable concentrations of salts have developed. Halite commonly forms thin surface crusts, and both gypsum and high

A

B

Figure 46. Friable red sandstone cross-beds forming *A*, interdune floor where bevelled at Soutrivier; *B*, residual hills and ledges bordering Tsondab Vlei.

Figure 47. Trench in sandy interdune at 6 km south of Rooibank showing bevelled dune structures underlying low-angle ripple laminae of floor; scale in inches. Photograph by E. Tad Nichols.

magnesian calcite may be present. These features have been described in some detail by Nagtegaal (1973, p. 5, 7, 8) and their distribution is mapped by him (1973, Fig. 1). Black muds are present in some coastal interdunes.

Adhesion ripples that are perhaps the most characteristic feature of interdune sebkhas are described by Glennie (1970) as structures formed where wind-blown sand adheres to the wet surfaces of mounds or ridges to build up small humps. Their size, as determined by Nagtegaal (1973, p. 7), ranges from 0.5 cm to 12 cm in cross-section. Mostly they are 2 to 6 cm in cross-section and 1 to 3 cm in height. They tend to form "a distinctive, irregular, lenticular bedding."

Varieties of adhesion structures have been recognized since the terms "adhesion ripple" and "adhesion wart" were introduced by Reineck (1955). Adhesion wart (Fig. 5) was suggested for surface bumps with lenticular, convex-up structures in cross-section. Adhesion laminae was adopted by Kocurek (in press) where sand streaming across a damp surface is trapped to form thin laminae. The common structures in the Namib Desert should probably be considered adhesion warts, according to Kocurek.

Adhesion structures commonly form on supratidal flats that merge into the barchan dune field. Thus, surface cementation with halite is important to their formation. These deposits are associated with dense brines derived from concentration of seawater, and they form precipitates of various salts in the interdunes. Deposits of adhesion-ripple sand, accumulated by wind deposition, are described (Nagtegaal, 1973, p. 7) as up to 50 cm high and resting on a layer of evenly spread beach sand across a wide area. Locally the adhesion ripples are eroded by wind into elongate forms.

Minor sedimentary structures of the sebkha interdunes include, in addition to various forms of the abundant adhesion structures, numerous features characteristic of water deposition, such as water ripple marks, contorted and brecciated beds resulting from sediment loading, scour channels, subaqueous cross-strata, and others. Thus, interdune deposits of this type should be easy to distinguish from those of the dry desert farther inland.

Interdune Lakes

Evidence of ancient lake deposits in various interdune areas of the Namib sand sea has been reported. At Narabeb, for example, 20 km south of the Kuiseb River and 47 km east of the Atlantic Ocean, the shoreline of a shallow lake is said by Selby and others (1979) to be indicated by precipitates of calcium carbonate. Around the borders of the lake, which is about 0.5 to 2 km in diameter, are found "scrapers and other human artifacts." Other possible lake deposits occur in a large interdune south of Gobabeb where early man sites have been found in an area of extensive calcrete deposits.

Acknowledgments

Field studies of the Namib Desert dune structures were conducted by a team of four people: E. D. McKee (principal investigator), Barbara McKee (field assistant), E. Tad Nichols (photographer), and Mary Jane Nichols (recorder). The investigation was made possible through the courtesy of M. K. Seely, Director of the Desert Ecological Research Unit of the Namib Desert Station at Gobabeb, near Walvis Bay, South West Africa. Other members of the research station staff, whose assistance in the field was much appreciated, are Peter Mostert, Ron Tilson, Frank Von Blottnitz, and Kate O'Brien. Thanks are extended to Peter Hugo, director, and Eugene Joubert of the CSIR (Council of Scientific and Industrial Research) who gave support and much encouragement to the project. For technical criticism and helpful suggestions following review of the manuscript, appreciative thanks are extended to the following: M. K. Seely, N. Lancaster, John Ward, Heim Tsoar, Gary Kocurek, T. S. Ahlbrandt, Christopher Schenk, Tad and Mary Jane Nichols. Drafting of sketches and charts was by William Chesser, except where otherwise noted.

References Cited

Ahlbrandt, T. S., Andrews, Sarah, and Gwynne, D. T., 1978, Biturbation in eolian deposits: Journal of Sedimentary Petrology, v. 48, p. 839–848.

Aufrere, Léon, 1928, L'orientation des dunes et la direction des vents: [Paris] Comptes Rendus Hebdomadaires des Séaucesde l'Académie des Sciences, v. 187, p. 833–835.

——1932, Morphologie dunaire et météorologie saharienne: Bulletin de l'Association de Géographes Français 56, p. 34–47.

——1934, Les dunes du Sahara algerien: Bulletin de l'Association Géographes Français 83, p. 130–142.

Bagnold, R. A., 1941, The physics of blown sand and desert dunes: New York, William Morrow and Company, 265 p.

——1953, The surface movement of blown sand in relation to meteorology, *in* Desert research—International Symposium, Jerusalem, 1952, Proceeding: Israel Research Council Special Publication 2, p. 89–96.

Barnard, W. S., 1973, Duinformasies in die Sentral Namib: Tegnikon, December, p. 2–13.

Besler, Helga, 1975, Messungen zur Mobiltat Dunensanden am Nordrand der Dune-Namib (Sudwestafrika): Würtzburger Geographische Arbeiteu, v. 43, p. 135–147.

——1976, Wasseruberformate Dune als Glied in der Landschaftsgenese der Namib: Communications from the Basal Africa Bibliography, v. 15, p. 83–106.

——1977, Untersuchungen in der Dunen-Namib (Sudwestafrika): South West Africa Wissenschaftliche Gesellschaft, Windhoek, South West Africa Journal 31, p. 33–64.

Besler, Helga, and Marker, M. E., 1979, Namib Sandstone: A distinct lithologic unit: Transactions of the Geological Society of South Africa, v. 82, p. 155–160.

Capot-Rey, Robert, 1947, L'Edeyen de Mourzouk: Algiers Université, travaux de l'Institut de Recherches Sahariennes, v. 4, p. 67–109.

Capot-Rey, Robert, and Capot-Rey, Françoise, 1948, Le déplacement des sables éoliens et la formation des dunes désertiques, d'aprés R. A. Bagnold: Algiers Université, Travaux de l'Institut de Recherches Sahariennes, v. 5, p. 47–80.

Clifford, T., 1967, The Damaran Episode in the upper Proterozoic–lower Paleozoic structural history of southern Africa: Geological Society of America Special Paper 92, 100 p.

Cooke, R. U., and Warren, Andrew, 1973, Geomorphology in deserts: Los Angeles, California University Press, 374 p.

Cooper, W. S., 1958, Coastal sand dunes of Oregon and Washington: Geological Society of America Memoir 72, 169 p.

——1967, Coastal dunes of California: Geological Society of America Memoir 104, 131 p.

Cornish, Vaughan, 1897, On the formation of sand dunes: Geographical Journal, v. 9, p. 278–309.

Fryberger, S. G., 1979, Dune forms and wind regime, *in* McKee, E. D., A study of global sand areas: U.S. Geological Survey Professional Paper 1052, p. 137–170.

Glennie, K. W., 1970, Desert sedimentary environments, *in* Developments in sedimentology [v.] 14: Amsterdam, London, New York, Elsevier Publishing Co., 222 p.

——1972, The Permian Rotliegendes of northwest Europe interpreted in light of modern desert sedimentation studies: American Association of Petroleum Geologists Bulletin, v. 56, no. 6, p. 1048–1071.

Goudie, A. S., 1972, Climate, weathering, crust formation, dunes and fluvial features of the Central Namib Desert, near Gobabeb, South West Africa: Madoqua, Ser. II, v. 1, nos. 54–64, p. 15–31.

Hanna, S. R., 1969, The formation of longitudinal sand dunes by large helical eddies in the atmosphere: Journal of Applied Meteorology, v. 8, p. 874–883.

Hedin, Sven, 1896, A journey through the Takla-Makan Desert, Chinese Turkistan: Geographical Journal, v. 8, July to December, p. 264–278.

——1905, The central Asian deserts, sand-dunes, and sands, Part 2, *in* Scientific results of a journey in central Asia, 1899–1902, Volume 2: Stockholm, Swedish Army Lithographic Institite, p. 379–718.

Holm, D. A., 1960, Desert geomorphology in the Arabian peninsula: Science, v. 132, p. 1369–1379.

Hunter, R. E., 1980, Quasi-planar adhesian stratification—An eolian structure formed in wet sand: Journal of Sedimentary Petrology, v. 50, p. 263–266.

Kaiser, E., 1926, Die Diamantenwüste Südwestafrikas: Berlin, v. 1, 321 p.; v. 2, 535 p.

King, D., 1960, The sand ridge deserts of South Australia and related aeolian landforms of Quaternary arid cycles: Royal Society of South Australia Transactions, v. 83, p. 99–108.

King, L. C., 1967, The morphology of the Earth (2nd ed.): Edinburgh, Oliver and Boyd, 699 p.

Koch, C., 1961, Some aspects of abundant life in the vegetationless sand of the Namib Desert dunes: Journal of the South West Africa Scientific Society, v. 15: p. 8–34, 77–92.

Kocurek, Gary, 1981, Significance of interdune deposits and bounding surfaces in eolian dune sands: Sedimentology (in press).

Lancaster, N., 1980, The formation of seif dunes from barchans—Supporting evidence for Bagnold's model from the Namib Desert: Zeitschrift für Geomorphologie, v. 24, no. 2, p. 160–167.

——1981, Grain size characteristics of Namib Desert linear dunes: Sedimentology, v. 28, p. 1–8.

Logan, R. F., 1960, The central Namib Desert, South West Africa: National Academy of Science, National Research Council Publication 758, 162 p.

Madigan, C. T., 1946, The Simpson Desert Expedition, 1939, Scientific reports; No. 6, Geology—The sand formations: Royal Society of South Australia Transactions, v. 70, p. 45–63, 205–227.

Mainguet, Monique, 1976, Propositions pour une nouvelle classification des édifices sableux éolian d'après les image des satellites Landsat I, Gémini, NOAA 3: Zeitschnift für Geomorphologie, v. 20, no. 3, p. 275–296.

McKee, E. D., 1945, Small scale structures in Coconino Sandstone of northern Arizona: Journal of Geology, v. 53, no. 5, p. 313–325.

——1957, Primary structures in some recent sediments: American Association of Petroleum Geologists Bulletin, v. 41, no. 8, p. 1704–1747.

——1966, Structures of dunes at White Sands National Monument, New Mexico: Sedimentology, v. 7, p. 1–60.

——1979, A study of global sand seas: U.S. Geological Survey Professional Paper 1052, 423 p.

McKee, E. D., and Tibbitts, C. G., Jr., 1964, Primary structures of a seif dune and associated deposits in Libya: Journal of Sedimentary Petrology, v. 34, no. 1, p. 5–17.

McKee, E. D., Douglass, J. R., and Rittenhouse, Suzanne, 1971, Deformation of lee-side laminae in eolian dunes: Geological Society of America Bulletin, v. 82, no. 2, p. 359–378.

Melton, F. A., 1940, A tentative classification of sand dunes—Its application to dune history in the southern High Plains: Journal of Geology, v. 48, no. 2, p. 113–174.

Monod, Théodore, 1958, Majabat al-Koubra; Contribution à l'étude de l'"empty-quarter" ouest-saharien: Français d'Afrique Noire Institute Memoire 52, 406 p.

Nagtegaal, P.J.C., 1973, Adhesion-ripple and barchan dune sands of the recent Namib (SW Africa) and Permian Rotliegend (NW Europe) deserts: Madoqua, ser. II, v. 2, nos. 63–68, p. 5–19.

Petrov, M. P., 1967, The Hoshi Corridor, Tsaidam and Tarim Basin, Volume 2 *of* The deserts of Central Asia: U.S. Joint Publications Research Service JPRS 42,772, 335 p. [translated from Russian].

Reineck, H. E., 1955, Haftrippeln und Haftwarzen, Ablagerungsformen von flugsand: Senckenbergiana Lethaea, v. 36, p. 347–357.

Royal Navy and South African Air Force, 1944, Introduction: The west coast of Africa from River Congo to Olifants River, with an appendix on the climates of St. Helens and Tristan da Cunha, Part 1, *in* Weather on the coasts of south Africa, v. 2: p. iii–viii, pt. 1, p. 1–61.

Rust, Uwe, and Wieneke, Friedrich, 1974, Studies on gramadulla formation in the middle part of the Kuiseb River, South West Africa: Madoqua, ser. II, v. 3, nos. 69–73, p. 5–15.

Scholz, H., 1968, Die Boden der Wuste Namib, Sudwestafrika: Zeitschrift für Pfanzenernahrung und Bodenkunde, v. 119, p. 91–107.

——1972, The soils of the central Namib Desert with special consideration of the soils in the vicinity of Gobabeb: Madoqua, ser. II, v. 1, nos. 54–62, p. 33–51.

Seely, M. K., and Sandelowsky, B. H., 1974, Dating the regression of a river's end point: South Africa Archeological Society Bulletin, Goodwin Series no. 2, p. 61–64.

Selby, M. J., 1976, Some thoughts on the geomorphology of the Central Namib Desert: Namib Bulletin, Supplement no. 1 to the Transvaal Museum Bulletin, August 1976, p. 5–6.

Selby, M. J., Hendy, C. H., and Seely, M. K., 1979, A late Quaternary lake in the central Namib Desert, southern Africa, and some implications: Paleogeography, Paleoclimatology, Paleoecology, v. 26, p. 27–41.

Smith, H.T.U., 1946, Sand dunes: New York Academy of Science Transactions, ser. 2, v. 8, no. 6, p. 197–199.

Spreitzer, H., 1965, Beobachtungen zur Geomorphologie der Zentralen Namib und ihrer Randgebiete, Sonder-ver offentlichung Windhoek: South West Africa Scientific Society, v. 4, 34 p.

Stone, R. O., 1967, A desert glossary: Earth-Science Review, v. 3, no. 4, p. 211–268.

van Straaten, L.M.J.U., 1953, Rhythmic pattern on Dutch North Sea beaches: Geologie en Mijnbouw, v. 15, p. 31–43.

Tsoar, Haim, 1978, The dynamics of longitudinal dunes—Final technical report: London, England, U.S. Army, European Research Office, 171 p. [Ben-Gurion University of the Negev, Department of Geography, Grant number DA-ERO 76-G-072.]

Verstappen, H. T., 1968, On the origin of longitudinal (seif) dunes: Zeitschrift für Annals Geomorphologie, v. 12, no. 2, p. 200–220.

Walker, T. R., 1979, Red color in dune sand, *in* McKee, E. D., A study of global sand seas: U.S. Geological Survey Professional Paper 1052, p. 61–82.

Wilson, I. G., 1971, Desert sandflow basins and a model for the development of ergs: Geographical Journal, v. 137, pt. 2, p. 180–199.

——1973, Ergs: Sedimentary Geology, v. 10, no. 2, p. 77–106.

Wopfner, H., and Twidale, C. R., 1967, Geomorphological history of the Lake Eyre Basin, *in* Jennings, J. N., and Mabbutt, J. A. (eds.), Landform studies from Australia and New Guinea: London, Cambridge University Press, p. 118–143.

MANUSCRIPT RECEIVED BY THE SOCIETY AUGUST 10, 1981
REVISED MANUSCRIPT RECEIVED AND ACCEPTED SEPTEMBER 25, 1981

Typeset by WESType Publishing Services, Inc., Boulder, Colorado
Printed in the U.S.A. by InterCollegiate Press, Inc., Shawnee Mission, Kansas